3rd International Workshop on Infrared Plasma Spectroscopy 2008

(IPS 2008)

Journal of Physics: Conference Series Volume 157

Greifswald, Germany
23-25 July 2008

ISBN: 978-1-61738-376-2

Printed from e-media with permission by:

Curran Associates, Inc.
57 Morehouse Lane
Red Hook, NY 12571
www.proceedings.com

Some format issues inherent in the e-media version may also appear in this print version.

Printed by Curran Associates, Inc. (2010)

Institute of Physics
Dirac House, Temple Back
Bristol BS1 6BE UK

Phone: 44 1 17 929 7481
Fax: 44 1 17 920 0979

techtracking@iop.org

TABLE OF CONTENTS

PREFACE

This volume containsd a selection of papers from the third Infrared Plasma Spectroscopy (IPS) Workshop held in Greifswald, Germany in July 2008. Although not all the contributions have been written up in time for the deadline for this volume, nevertheless the 12 contributions presented here give a fair representation of the conference topics. The conference comprised four different types of contribution. Firstly, four invited lectures focussed on the prime areas of interest. Secondly, eight shorter contributed talks, grouped as closely as possible with the appropriate invited lecture. These contributed talks covered topics in both pure and applied infrared plasma spectroscopy. A feature of the two previous IPS conferences has been a contribution from commercial organisations namely those involved in manufacturing devices, detectors and spectrometers. This group of participants formed the third part of the conference programme and gave five oral presentations covering topics like QCL and detector/detection developments and novel spectrometer designs. The fourth contributing group comprised 27 poster presentations. It should be mentioned that some of the latter were poster versions of contributed talks. The conference was remarkable for the wide spread of topics covered in a relatively small meeting, consisting of 44 participants. The participants were made up of 34 scientists from within Europe and 4 from the rest of the world. It is interesting to reflect on changes that have occurred since the previous meeting just a year earlier. Two clear developments which have occurred are the emergence of Quantum Cascade Lasers (QCL) and their use in Cavity Ring Down (CRD) spectroscopy. A major shift from cw lead salt diode lasers to cw and pulsed QCL in both pure and applied projects now seems to be well under way. The topics covered in the earlier conferences focussed more on applying infrared spectroscopy to plasma monitoring and control. When choosing the topics to cover the scientific committee felt that this time it would be useful to emphasise new spectroscopic developments as well as covering applications. This might serve as a guide as to where the subject of infrared spectroscopy in combination with plasma sources might be heading in the future i.e. to emphasize pure infrared spectroscopy developments. The first invited lecture (G Guelachvili and N Picque) and the last invited lecture (F K Tittel, Y Bakhirkin, R Curl, A Kosterev, R Lewicki, D Thomasz and S So) were chosen to set the scene and realise this objective. The second (R Engeln, R Zijlmans, S Welzel, O Gabriel, J-P van Helden, J Röpcke and D Schram) and third (X Aubert, C Lazzaroni, D Marinov, O Guaitella, S Welzel, A Pipa, J Röpcke and A Rousseau) invited talks focussed on the application of the IR laser techniques with particular emphasis on the role of surfaces in plasmas and the relevance of plasma surface interactions. Surface plasma interactions did not feature strongly in the two earlier meetings and so this topic too, along with the emphasis on novel infrared spectroscopy techniques, represents a new direction for the conference.

Paul B Davies and Jürgen Röpcke

International Scientific Committee

P B Davies, Cambridge, UK: Chair
J Röpcke, Greifswald, Germany: Co-Chair
R Engeln, Eindhoven, Netherlands
G Hancock, Oxford, U K
M Hori, Nagoya, Japan
H Linnartz, Leiden, Netherlands
R Martini, New York, USA
J Meichsner, Greifswald, Germany
A Rousseau, Paris, France

Local Organizing Committee

J Röpcke (INP: Chair)
F Hempel (INP: Secretary)
J Meichsner (IfP, University of Greifswald)
N Lang (INP)
L Glawe (INP)
C Krcka (INP)
B Lindemann (INP)

FTIR and QCL diagnostics of the decomposition of volatile organic compounds in an atmospheric pressure dielectric packed bed plasma reactor

Z Abd Allah[1,3], **D Sawtell**[1], **V L Kasyutich**[2], **P A Martin**[1]

[1] School of Chemical Engineering and Analytical Science, University of Manchester, Sackville Street, Manchester, M60 1QD, UK
[2] TDL Sensors Ltd., CTF, 46 Grafton St., Manchester, UK
[3] Author to whom corrspondance should be addressed

E-mail: zaenab.abdallah@postgrad.manchester.ac.uk

Abstract. Non-thermal atmospheric plasmas can be used to decompose toxic volatile organics. A dielectric packed bed plasma reactor was used to decompose dichloromethane (DCM) in a flowing gas mixture of nitrogen and oxygen. The effect of varying the oxygen concentration on the decomposition of dichloromethane was investigated. A comparison has been made of spectroscopic diagnostics based on FTIR (Fourier Transform Infrared) spectroscopy and infrared quantum cascade laser spectroscopy for the detection of nitric oxide produced in the plasma reactor.

1. Introduction

Volatile organic compounds which have been released to the atmosphere via vehicle exhausts, petroleum storage and distribution, oil refining, surface coating and other industrial processes contribute significantly to both air pollution and human health problems[1]. Some of these VOCs affect the human senses through their odour. Others have much more serious effects such as reducing the birthweight or even causing abortion. Many VOCs are also carcinogens.

Non-thermal plasma has been shown to have a great potential for the decomposition of VOCs. These are non-equilibrium plasmas, in which the energy is delivered to the electrons and not to the translational and rotational energy of the gas molecules thus requiring less input energy. These energetic electrons are responsible for initiating the chemical reactions in the plasma. They can also be generated at atmospheric pressure and room temperature which gives them an advantage for applications such as air pollution reduction[2, 3]. There are a variety of non-thermal plasma reactors such as, dielectric-barrier plasma reactors[4], pulsed corona plasma reactors[5], microwave plasma reactors[6], surface discharge reactors[7] and packed bed plasma reactors[8, 9]. The use of a packed bed reactor provides the advantages of uniform distribution of gas flow and discharge in the reactor. It can also be easily modified to incorporate a catalyst. Indeed the combination between catalyst and packed bed could reduce the amount of some by-products like NO_x, CO_2 and ozone[10].

In this work, non-thermal plasma generated in packed bed plasma reactor at atmospheric pressure is used to decompose dichloromethane (DCM) as an example of a chlorinated VOC. There have been several previous studies of the decomposition of DCM[5, 9-11]. Experiments have been conducted to investigate the effect of the addition of oxygen on the decomposition of dichloromethane. For diagnostics of the chemical species in the plasma, in-line FTIR (Fourier Transform Infrared) spectroscopy and mid-infrared quantum cascade laser spectroscopy are used. A comparison is then made of these two complementary techniques.

2. Experimental description

2.1. Packed bed plasma reactor system

Figure 1 shows a sketch of the experimental set-up. The packed bed reactor was constructed using a Pyrex tube of 20 cm length and 31 mm internal diameter and containing two stainless steel electrodes made with funnel ends separated by about 3 cm. A stainless steel mesh was placed in front of the electrodes and the volume filled with barium titanate ($BaTiO_3$) beads. These have a dielectric constant of 2000-10000.The packed bed volume was 24.9 cm^3, whilst the volume of the beads was about 9 cm^3 and the void volume is about 15.9 cm^3. For a 1 L min^{-1} flow rate a residence time of 0.95 s was achieved. The gas mixture passed into the reactor through a 1/4" tube in the centre of the electrodes. The inlet system has been designed to allow the sampling of the gas stream before and after the plasma reactor. In these experiments nitrogen was used as the carrier gas. Air and oxygen cylinders have been used to provide different percentages of oxygen to the plasma. The gases were used as supplied and consisted of an air cylinder (79 % nitrogen, 21 % oxygen), and a nitrogen cylinder (oxygen free) (99.998 % purity). A mass balance calculation allowed the adjustment of the gas flow through mass flow controllers (MKS 247). A total flow rate of 1 L min^{-1} including approximately 500 ppm dichloromethane was used. To obtain 500 ppm, dichloromethane (99.6 %) was cooled down to -20 °C using a coolant of 50 % water and 50 % antifreeze. A neon sign power supply (Hansen) was used to generate a high voltage applied to the reactor electrods. Electrical measurements of the voltage was made using a Tektronix 1000 to 1 voltage probe and the current was measured using a Pearson 411 current monitor. Plasma is generated in the packed bed reactor at atmospheric pressure and close to room temperature.

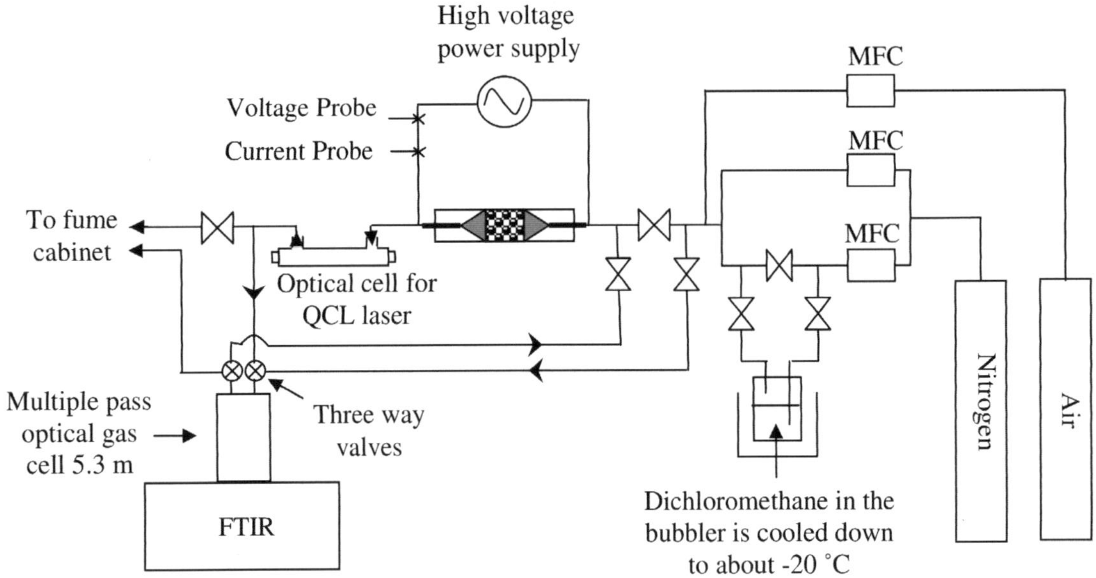

Figure 1. Sketch illustrating the experimental set up.

2.2. Diagnostics with in-line FTIR Spectroscopy:

FTIR spectroscopy (Bruker Equinox 55) was used to identify and determine the concentration of components in the reactor inlet and outlet gas streams. The gas streams were passed alternately into a multiple pass optical gas cell (Specac Ltd) with a White-type mirror arrangement producing an optical pathlength of 5.3 meters. Sampling was carried out about 0.4 sec after the packed bed reactor. FTIR analyses have been obtained at room temperature and atmospheric pressure at a spectral resolution of 2 cm^{-1}. A liquid nitrogen cooled MCT (mercury cadmium telluride) detector was also used for enhanced sensitivity. Concentrations of each species were determined by integrating the area under each peak using Bruker's OPUS software and then comparing with standard spectra produced by Pacific Northwest National Laboratories[12] at atmospheric pressure, one meter path length and different temperatures. The destruction efficiency of dichloromethane was calculated as follows:

$$\eta = [(C_{in} - C_{out}) / C_{in}] * 100\,\%$$

where C_{in} and C_{out} are the inlet and outlet concentration of dichloromethane respectively

2.3. Mid-infrared quantum cascade laser measurements of plasma products:

For fast response, high sensitivity and minimal spectral interferences, a thermoelectrically cooled, continuous wave (cw) distributed feedback (DFB) quantum cascade laser (QCL) spectrometer (TDL Sensors Ltd) was used to measure nitric oxide. The wavelength of the laser was current-tuned over two absorption peaks at ~1900.52 cm^{-1} and ~1900.075 cm^{-1} at a QCL temperature of -6.5 °C[13]. The plasma exhaust was connected to an absorption cell of pathlength 21 cm at atmospheric pressure. The exhaust then passes to the multiple pass cell used by the FTIR spectrometer as seen in figure 1. The detector used in this arrangement was a thermoelectrically cooled MCT detector. Figure 2 shows an example spectrum for measuring nitric oxide concentration at the exhaust of the plasma reactor. The inlet gas consists of approximately 500 ppm dichloromethane and 21 % added oxygen. It can be seen that the first 20 channel points are for no current passing through the laser in order to determine the 0% absorption level. The red line is a non-linear least squares fit to lorentzian lineshapes carried out in real-time to extract the concentration. Each spectrum is the result of 1000 averages in approximately 1 s.

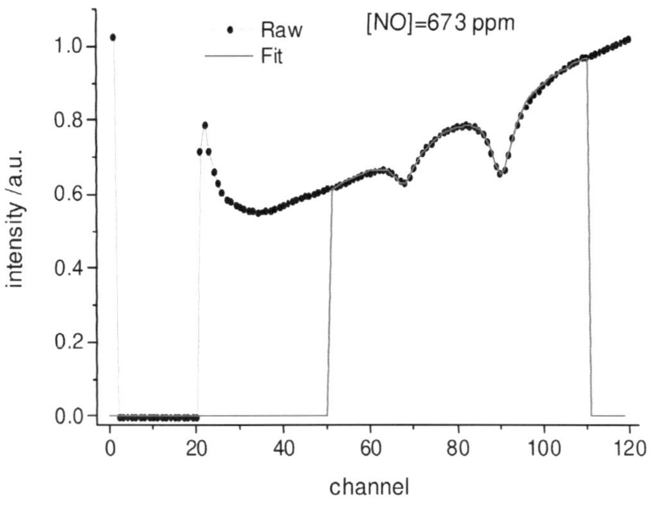

Figure 2. Raw signal and fit in plasma measurements at 300 K and atmospheric pressure. DCM=500 ppm, O_2=21 %, 0.5 L min^{-1}.

3. Results

3.1. Effect of adding oxygen to the plasma:

Figure 3 shows the FTIR spectrum, a) before and, b) after the reactor containing a nitrogen plasma with 500 ppm dichloromethane at a flow rate of 1 $L.min^{-1}$. Oxygen was then added to the gas stream by addition of air to the nitrogen plasma. Oxygen percentages were varied between 0 to 21 percent. Figure 4 shows a FTIR spectrum taken after plasma reactor for a gas stream of approximately 500 ppm DCM and 3 % added oxygen in nitrogen plasma. A maximum destruction of about 75 % at 3 % added oxygen to the gas stream has been found with nitrogen plasma.

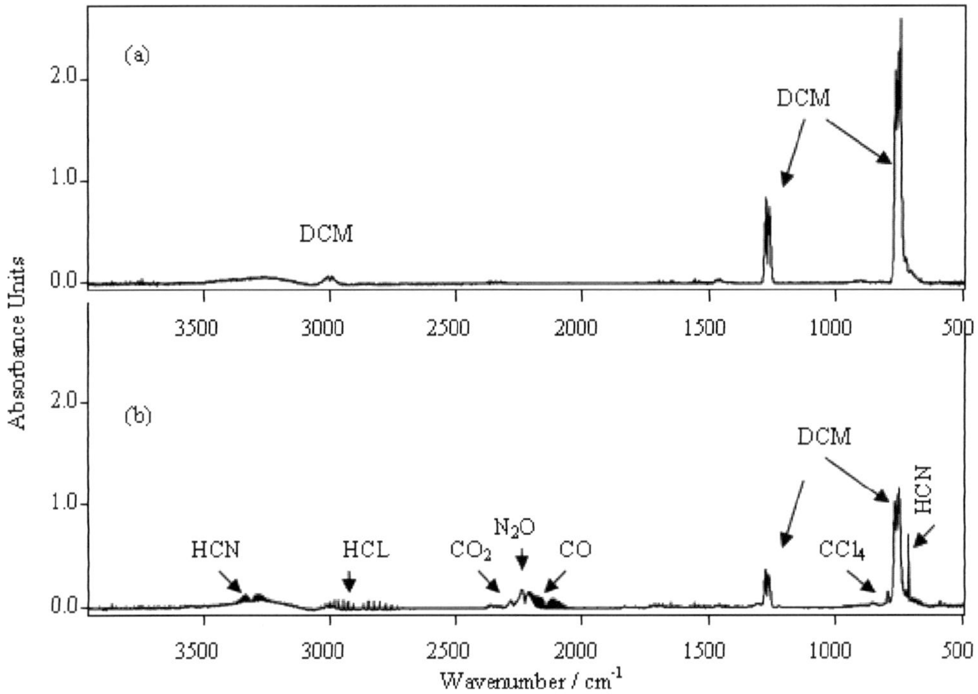

Figure 3. Infrared spectrum for 500 ppm dichloromethane in pure nitrogen with flow rate of 1 $L.min^{-1}$ (a) before the plasma reactor. (b) after the plasma reactor.

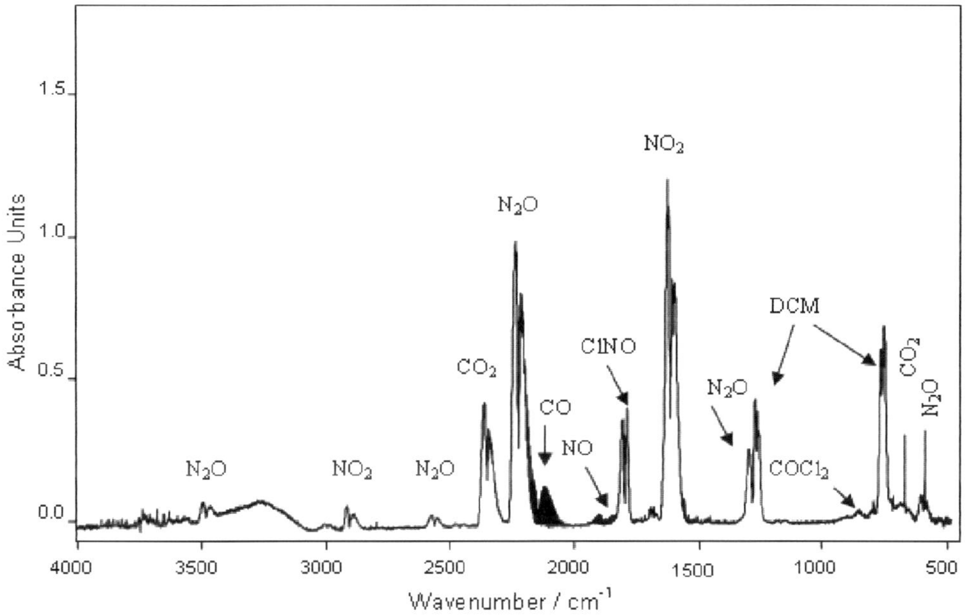

Figure 4. FTIR spectrum for approximately 500 ppm dichloromethane in 97 % nitrogen and 3 % oxygen after the plasma reactor.

The destruction of dichloromethane as well as the concentration for all species present in the examined plasmas with different percentages of added oxygen in nitrogen is shown in table 1. Phosgene ($COCl_2$), carbon tetrachloride (CCl_4) and hydrogen chloride (HCl) were not detected after adding oxygen to the gas stream, whilst HCN concentration was below the limit of detection at oxygen concentration higher than 1%.

Table 1. Decomposition efficiency for dichloromethane and the concentrations of all species detected in the plasma exhaust by FTIR as a function of oxygen concentration in nitrogen plasma.

	Destruction	Concentration / ppm										
O_2%	DCM%	DCM	ClNO	CO	CO_2	N_2O	NO_2	NO	CCl_4	HCl	HCN	$COCl_2$
0	53	261	1	77	2	23	0	0	4	68	50	0.4
1	67	183	4	117	19	117	63	17	.	.	3	.
2	74	142	11	113	24	136	105	33	.	.	.	.
3	75	142	17	118	28	146	147	49	.	.	.	.
4	72	155	19	109	33	148	169	59	.	.	.	.
5	70	168	22	108	37	143	189	74	.	.	.	.
7	70	169	27	110	48	156	225	133	.	.	.	.
9	66	190	28	108	58	151	255	155	.	.	.	.
10	58	235	29	114	63	143	262	151	.	.	.	.
13	57	241	32	116	77	142	306	195	.	.	.	.
15	55	250	34	120	87	147	339	227	.	.	.	.
17	50	278	36	124	98	143	344	234	.	.	.	.
21	54	257	26	99	102	113	294	193	.	.	.	.

The maximum destruction of dichloromethane in a nitrogen plasma occurs at 3 % added oxygen. This can be seen in figure 5 where the concentrations of the byproducts: nitrosyl chloride (ClNO), carbon monoxide (CO) and carbon oxide (CO_2) in nitrogen plasma are also shown as a function of oxygen concentration.

The effect of oxygen concentration on the formation of nitric oxide has been simultaneously studied by both FTIR and infrared quantum cascade laser absorption spectroscopy. Figure 6 shows the results of two repeated experiments. The measured NO concentrations were very similar up to an oxygen concentration of 10%. After this there is a slight divergence of the QCL and FTIR measurements. This may be due to further oxidation of NO in the exhaust stream in between the QCL sample cell and the FTIR sample cell. Also shown are the DCM concentrations which again show the minimum level of DCM in the exhaust achieved for an oxygen concentration of around 3%.

Figure 7 shows the results of monitoring the NO concentration over time with the QCL for an oxygen concentration of 21% and then for 10%. Step changes of about 50 ppm in the concentrations of NO occur every five minutes. This effect had not previously been observed with the FTIR. The origin is thought to be due to slight changes in the injected power into the plasma. Measurements of the voltage and current have been made using a 1000 to 1 high voltage probe and a Pearson 411 current monitor connected to a digital oscilloscope. Typical measurements are shown in figure 8. Here it can be seen that the power supply generates pulses with a repetition of 100 pulses per second and a pulse duration of 3.25 ms. The peak to peak voltage is about 14 kV at a frequency of 20 kHz during the pulse. An average deposited power of about 15 W, with an energy density of 900 J L^{-1} was calculated where the energy density was calculated as a ratio of the deposited power to the gas flow rate. Closer analysis of the voltage and current waveforms reveals a phase change in the leading edge of the voltage pulse every 5 minutes or so, however further investigation is required to ascertain the effect of this on NO formation.

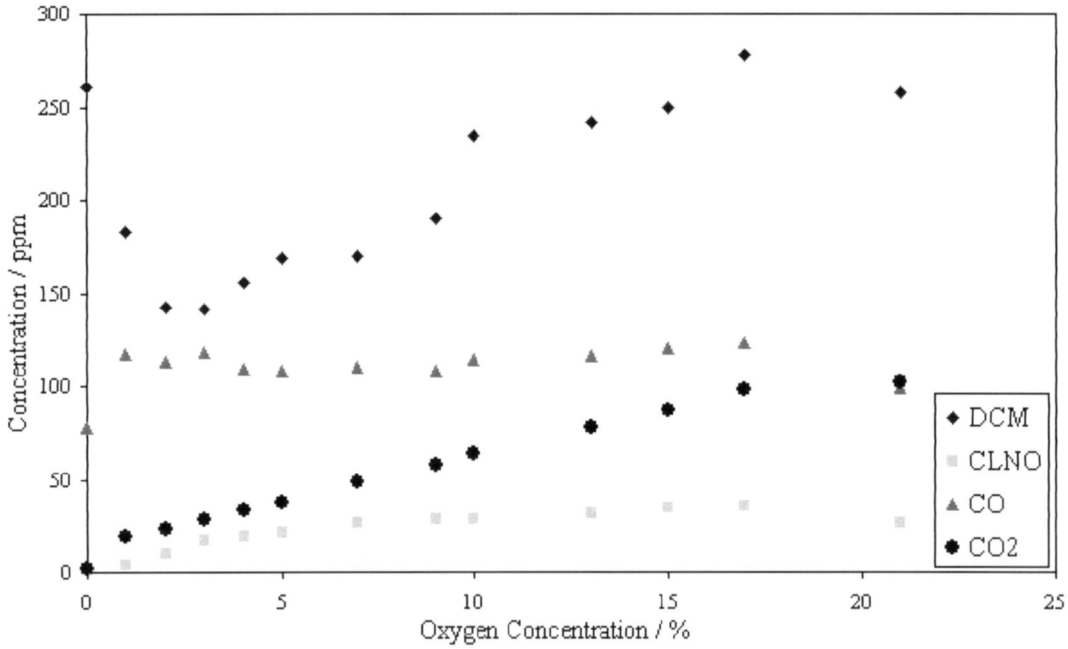

Figure 5. Comparison among the concentrations of dichloromethane (DCM), carbon monoxide (CO), carbon oxide (CO_2) and nitrosyl chloride (ClNO), all the concentrations are in ppm.

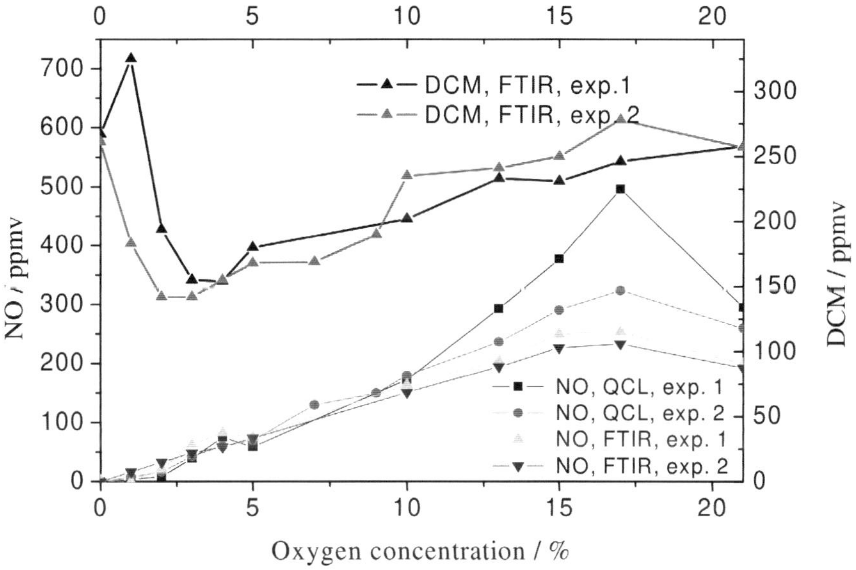

Figure 6. NO and DCM concentration versus percentage of oxygen in nitrogen in the plasma reactor with a flow rate of 1 $L.min^{-1}$ and a power of ~15 W. The initial DCM concentration was ~500 ppmv. QCL spectrometer with a 21 cm optical cell, V=103 cm^3 (t=6.1 s). FTIR with an 5.3 m optical White cell, V=1300 cm^3 (t=78 s).

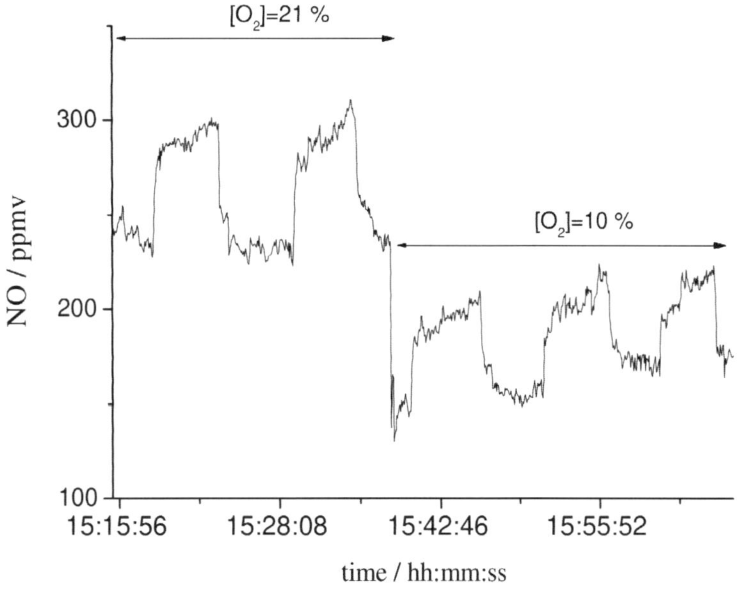

Figure 7. Nitric oxide concentration over time with 21 % and 10 % of added oxygen in nitrogen plasma with approximately 500 ppm DCM and total flow rate of 1 L min^{-1} measured via QCL laser spectroscopy.

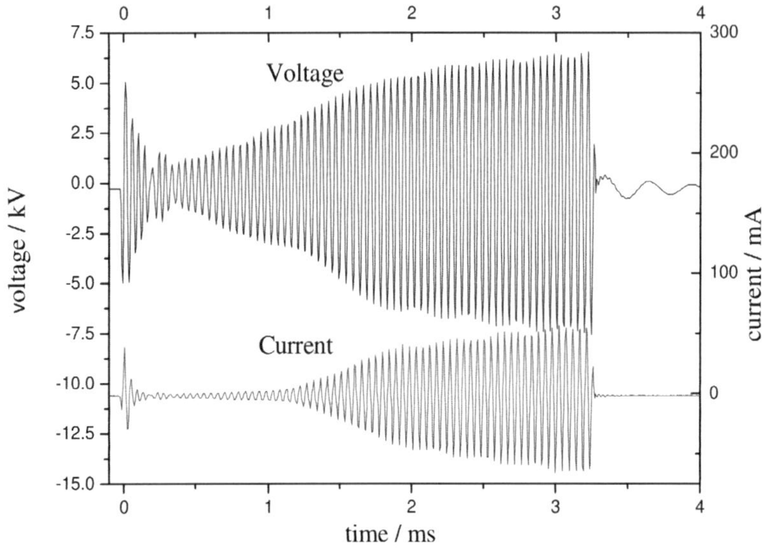

Figure 8. The voltage and current pulse applied to the electrodes.

4. Discussion and conclusions

This work has shown that the optimum destruction of DCM in nitrogen at atmospheric pressure in a dielectric packed bed reactor occurred at an oxygen concentration of 3%.This is in general agreement with Fitzsimmons et al[9] . At higher oxygen concentrations the formation of ozone becomes more likely which is less reactive than atomic O and OH in the destruction process. When there is no oxygen present the chlorine forms phosgene ($COCl_2$), carbon tetrachloride (CCl_4) and hydrogen chloride (HCl), whilst they were substantially reduced or not detected after adding oxygen.

The simultaneous measurement of NO by QCL and FTIR spectroscopy demonstrated the complementary nature of the two techniques. The laser approach provides a very rapid and highly sensitive measurement technique but for only a single species in this case whereas the FTIR has broad spectral coverage for multispecies detection but is less sensitive. It was encouraging to observe that the same NO concentrations were derived except for the situation of high oxygen concentration where the differing gas cell volumes and positions led to different measured NO concentrations. The QCL measurements of NO also showed up some instability in the power supply that was not detected by FTIR.

Further work is being directed towards direct in situ measurements with the QCL and FTIR with time-resolved detection on the microsecond timescale. Additionally multiple packed beds are also being explored and the addition of catalyst to enhance the conversion.

5. Acknowledgements

We would like to thank Andrew Campen for helpful discussions. Thank you also to John Riley, Loris Doyle, Paul Rothwell for technical assistance and to TDL Sensors Ltd for the loan of the QCL system. Zaenab Abd Allah would also like to thank the Capacity Building Project of the British Council and the Ministry of Higher Education in Syria for a studentship.

6. References

[1] Friedrich R and Reis S 2004 *Emissions of Air Pollutants*: Springer
[2] McAdams R 2001 *J.Phys. D* **34** 2810
[3] Oda T 2003 *Electrostatics* **57** 293

Journal of Physics: Conference Series **157** (2009) 012001 doi:10.1088/1742-6596/157/1/012001

[4] Kogelschatz U 2003 *Plasma Chem. and Plasma Proc* **23** 1
[5] Huang L, Nakajo K, Ozawa S and Matsuda H 2001 **35** 1276
[6] Rousseau A, Dantier A, Gatilova L, Ionikh Y, Röpcke J and Tolmachev Y 2005 *Plasma Sources Sci. Tech.* **14** 70
[7] Norberg A 1992 *IEEE Trans. Ind. Appl.* **28** 498
[8] Chen H L, Lee H M, Chen S H and Chang M B 2008 *Ind. & Eng Chem. Res.* **47** 2122
[9] Fitzsimmons C, Ismail F, Whitehead J C and Wilman J J 2000 *J. Phys. Chem. A* **104** 6032
[10] Wallis A E, Whitehead J C and Zhang K 2007 *Appl. Cat. B* **72** 282
[11] Li C T, Lee W J, Chen C Y and Wang Y T 1996 *J. Chem. Techn. Biotech* **66** 382
[12] Sharpe S W, Johnson T J, Sams R L, Chu P M, Rhoderick G C and Johnson P A 2004 *Appl. Spectroscopy* **58** 1452
[13] Kasyutich V, Holdsworth R and Martin P A 2008 *Appl. Phys. B* **92** 271

Journal of Physics: Conference Series **157** (2009) 012002 doi:10.1088/1742-6596/157/1/012002

A Mid-infrared QEPAS sensor device for TATP detection

C Bauer[1], U Willer[1], R Lewicki[2], A Pohlkötter[1], A Kosterev[2], D Kosynkin[3], F K Tittel[2], W Schade[1]

[1] TU Clausthal, Laser Application Centre, Am Stollen 19, 38640 Goslar, Germany, Phone: +49 5321 6855-186

[2] Electrical & Computer Engineering Department, Rice University, 6100 Main Street, MS-366 Houston, TX 77005, US

[3] Department of Chemistry, Rice University, 6100 Main Street, MS-222, Houston, TX 77005, US

E-mail: w.schade@pe.tu-clausthal.de, fkt@rice.edu

Abstract. Recent developments of external cavity quantum cascade lasers (EC-QC lasers) enable new applications in laser spectroscopy of trace gas species in the mid-infrared spectral region. We report the application of quartz enhanced photo acoustic spectroscopy (QEPAS) with widely tuneable EC-QC lasers as excitation sources for chemical sensing of different species such as triacetone triperoxide (TATP). A pulsed EC-QC laser operating at $\nu \sim 1120 cm^{-1}$ and a cw EC-QC laser operating at $\nu \sim 950 cm^{-1}$ are used for the detection of the explosive TATP which is a mid infrared broad band absorber. The detection limit of our present setup is ~1ppm TATP at atmospheric pressure.

1. Introduction

The use of an EC-QC laser for sensing broad band absorbers in the mid-infrared spectral region opens new possibilities in laser spectroscopy. Explosives have strong absorption features in the mid-infrared region especially between λ=7μm and λ=11μm. As an example a vapor phase FTIR-spectrum of TATP is given in figure 1. Different laser based methods have been used for the detection of explosives and the mid-infrared spectral region offers several possibilities for their sensing [1,2]. Since recent developments on QC laser technology this spectral region is covered by EC-QC lasers [3-5]. Recently a cw EC-QC laser with a tuning range covering $182 cm^{-1}$ combined with the output power of P=20mW for sensing different species was realized by Wysocki et al. [6]. In our sensing scheme we use two different commercially available EC-QC lasers: a pulsed laser operating between $1155 cm^{-1}$ and $1220 cm^{-1}$ and a cw laser operating between $925 cm^{-1}$ and $975 cm^{-1}$. The tuning ranges of both systems are marked in the TATP spectrum in the figure 1 as a reference. QEPAS has been used extensively in order to detect different species in the near infrared (NIR) spectral region and also in the mid-infrared region. Details of this technique are given in [7, 8]. The sensing of broad band absorbers (for example acetone) with QEPAS was also reported earlier by Lewicki [9]. TATP is a peroxide based explosive which is also a broad band absorber in the mid-infrared region [10]. In this report we present for the first time a QEPAS based sensor approach with a QCL for the detection of TATP.

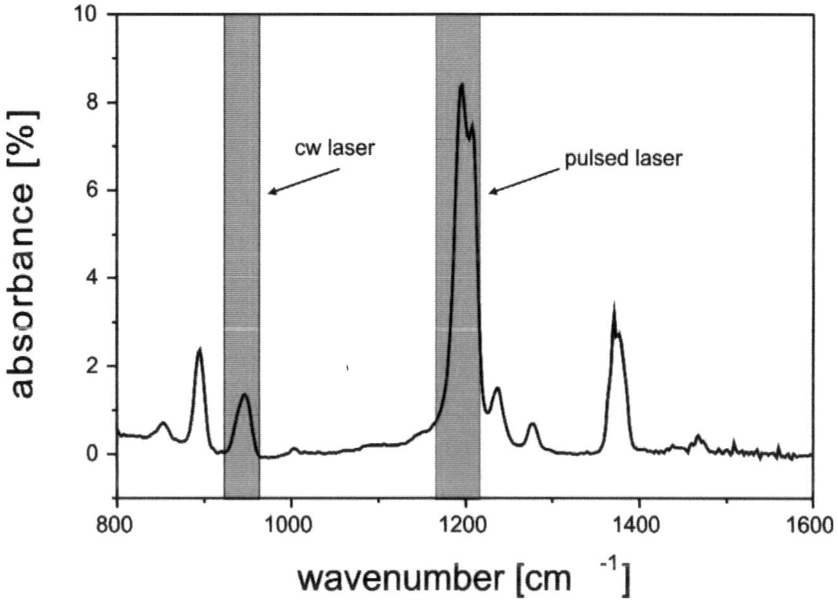

Figure 1. FTIR vapor spectrum of TATP and the tuning ranges of the used EC-QC lasers.

2. Experimental Setup

2.1. Quartz Enhanced Photo Acoustic Spectroscopy (QEPAS)

QEPAS has proven its applicability in sensing of different species. In this report we only give a brief introduction to this technique. The modulated laserlight is focused between two prongs of a quartz tuning fork (TF). Due to the vibration-translation energy transfer in the target molecule an acoustic wave is generated. If the laserlight is modulated at the resonance frequency of the TF a current signal can be measured due to the piezoelectric properties of the quartz material. The generated signal depends largely on the concentration and the absorption coefficient of the molecule and the optical power of the laser. For signal to noise (SNR) enhancement an acoustic micro-resonator (figure 2a) is extremely useful. As reported earlier the detection sensitivity can be increased if a micro resonator is used (for details see [11]). In our setup a QEPAS cell with two tubes (diameter: 600μm, length: 5mm) acting as an acoustic micro resonator is used (figure 2b-2d). The acoustic micro resonator is increasing the SNR by a factor ~7. The laserlight is coupled through a 1mm thick zinc selenide window into the cell.

Journal of Physics: Conference Series **157** (2009) 012002 doi:10.1088/1742-6596/157/1/012002

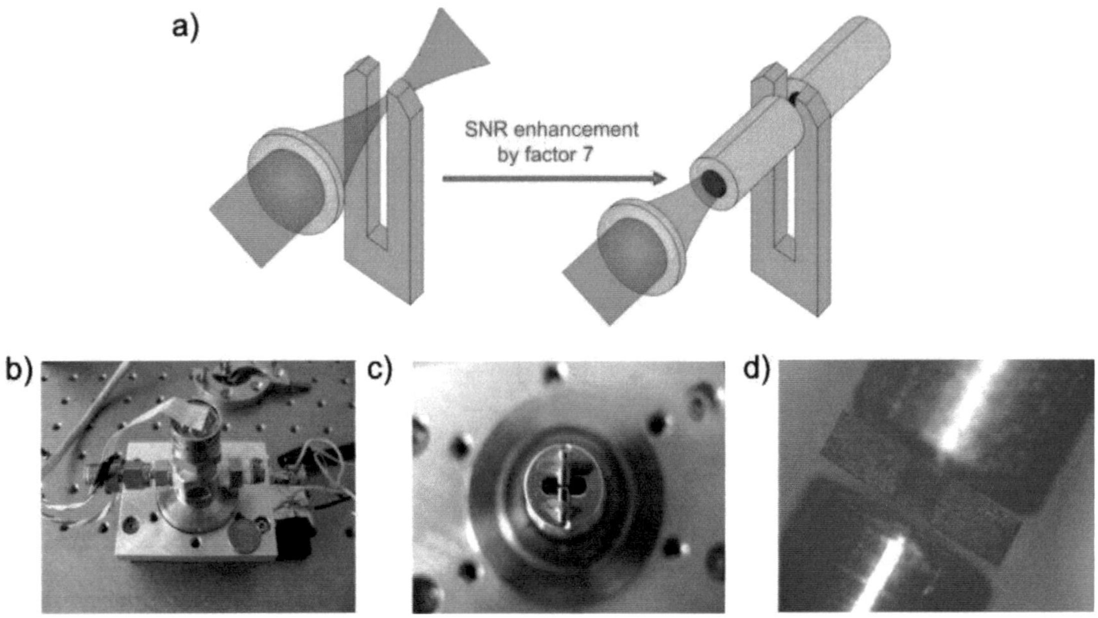

Figure 2. A QEPAS cell for sensing different species in the mid-infrared region: (a) principle of QEPAS and SNR enhancement with an acoustic micro-resonator. (b) The cell with zinc selenide windows. (c) The acoustic micro-resonator consists of two small tubes. (d) The quartz TF is placed between the tubes.

2.2. External Cavity Quantum Cascade Laser (EC-QC laser)

A fabry perot QC laser with AR-coated end facets is used as a gain medium with an external cavity in littrow configuration. This combination results in a widely tunable mid-infrared laser source with a high output power.

In this work two commercially available external cavity lasers (daylight solutions) are used. The laser housing is compact and in case of a pulsed laser no additional water cooling is needed. The tuning range of the pulsed laser (λ=8.4μm) is around $\Delta\nu$=75cm^{-1} (figure 3b) while the mode hop free cw laser (λ=10.5μm) can be tuned $\Delta\nu$=60cm^{-1} (figure 3c).

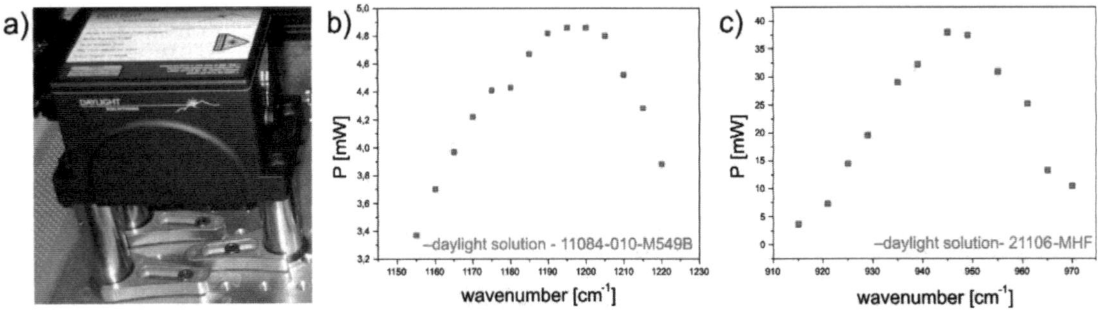

Figure 3. (a) laser housing of the EC-QC laser. (b) Tuning range of the pulsed laser at λ=8.4μm. (c) Tuning range of the mode hop free cw EC-QCL at λ=10.5μm.

For QEPAS a modulation of the laserlight at the resonance frequency of the TF is needed as well as the capability of scanning the wavelength. For coarse tuning of the wavelength the scanning option of the EC QC lasers can be used. A good agreement between measured emission wavelength and reading on the laser driver was found. For fine tuning, a piezo at the laserhead can be used. For calibration purposes of the used QEPAS cell different spectra of NH_3 were taken. For example a spectra of NH_3 with a concentration of 100ppm and at an ambient pressure of p=9torr is given in figure 4. A fine tuning coefficient of $6.75 \cdot 10^{-3}$ cm^{-1}/V_{pp} is derived from these measurements. A single absorption line at ν~932.88cm^{-1} was recorded also in order to show the small linewidth of the cw laser (figure 4b). The linewidth of the cw EC-QC laser is Δν <0.001cm^{-1}.

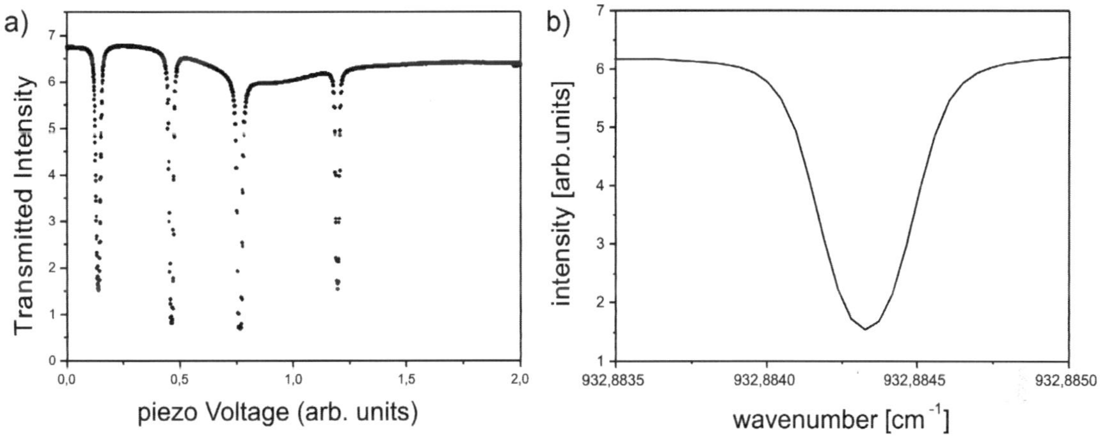

Figure 4. (a) "Piezo scan" for high resolution spectroscopy of ammonia at ν~932cm^{-1}. (b) A resolved single absorption line of ammonia at ν~932.88cm^{-1}.

For modulation of the laserlight three different approaches are used:

(1) amplitude modulation by application of an external trigger signal at f=32.756kHz. This is only possible with the pulsed EC-QC laser.
(2) amplitude modulation by use of a second tuning fork. The second tuning fork is placed into the beampath and driven into oscillation by application of an external voltage. Due to the vibrations of the prongs, the TF acts like a chopper. This method is used for the amplitude modulation of the cw EC-QC laser.
(3) frequency modulation for 2f-WMS. This method can only be applied for narrow linewidth absorbers and is therefore not suitable for broad band absorbers like TATP.

3. Results

The pulsed laser system (λ=8.4μm) was used for sensing TATP with a QEPAS sensor and for amplitude modulation the laser was triggered at f=32.756kHz at a duty cycle of 5% which results in an output power of P=5mW. The TATP was synthesized as described by Wolfenstein [12]. A small amount of TATP (<500μg) was placed inside the cell. TATP has an extremely high vapor pressure that corresponds to an equilibrium concentration of 65ppm at ambient air [13] and after a short time (few seconds) the TATP could be detected via QEPAS. Figure 5 shows the recorded TATP spectrum at ambient air. The two C-O stretch bands near ν=1200cm^{-1} can clearly be seen. Assuming the equilibrium concentration of ~65ppm for TATP at ambient air a theoretical detection limit of 1ppm TATP can be estimated with this setup.
For sensing the O-O stretch band of TATP at ν=945cm^{-1} the cw EC-QC laser was used. For amplitude modulation a second "chopper" TF was used. The SNR of the QEPAS signal is not as high as for the previously described measurement because of the lower absorption coefficient and due to the fact, that the modulation depth of the signal was only ~70%.

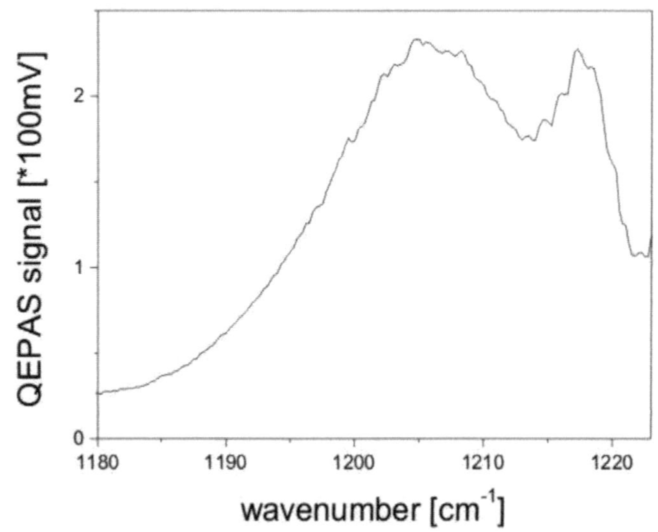

Figure 5. QEPAS signal from the complete C-O stretch band of TATP at around λ=8.4µm.

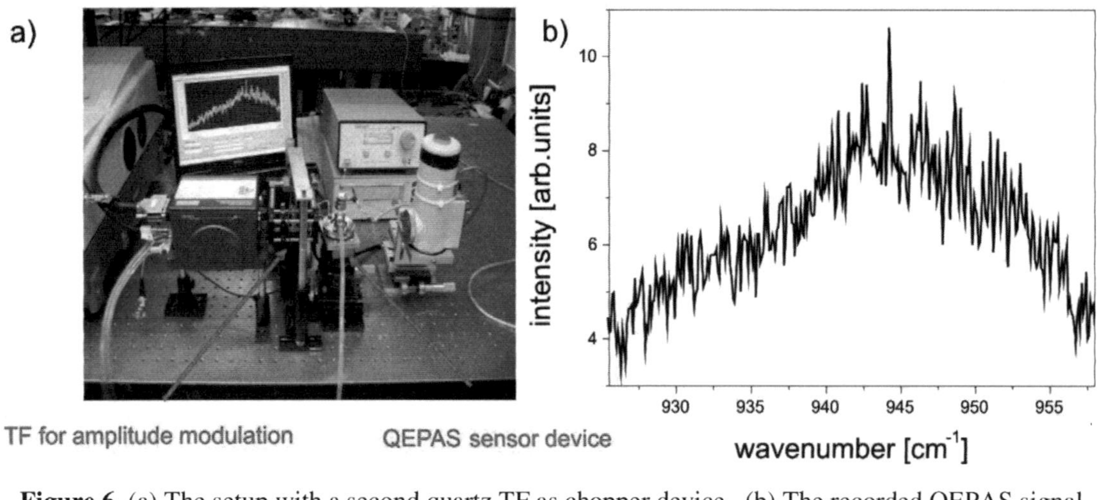

Figure 6. (a) The setup with a second quartz TF as chopper device. (b) The recorded QEPAS signal from the O-O band of TATP at around λ=10.5µm.

4. Conclusion and outlook

In this report we present a QEPAS sensor for the detection of the explosive TATP. Two different widely tunable EC-QC lasers are used. A pulsed one at λ=8.4µm in combination with a QEPAS cell enables a detection limit of 1ppm TATP in ambient air. Next steps are the integration of a QEPAS based sensor setup inside an existing security system (e.g. interlocks at airports) and the development of a fiber coupled QEPAS-TATP sensor system. With a mode hop free cw EC-QC laser it is possible

Journal of Physics: Conference Series **157** (2009) 012002 doi:10.1088/1742-6596/157/1/012002

to sense broad band absorbers like TATP but also small molecules with a resolved spectrum like NH_3. Narrow linewidth, high output power and wide tuning range are achieved with a mode hop free EC-QC laser. The setup of a multi species sensor system for the detection of several explosives is one of the upcoming challenges in homeland security and in our opinion mid-infrared spectroscopy will be one key technology in a multi sensor system.

References

[1] Todd M W, Provencal R A, Owano T G,Paldus B A, Kachanov A, Vodopyanov K L, Hunter H, Coy S L, Steinfeld J I and Arnold J T 2002 Appl. Phys. B **75** 367
[2] Bauer C, Sharma A K, Willer U, Burgmeier J, Braunschweig B, Schade W, Blaser S, Hvozdara L, Müller A, Holl G, 2008 Appl. Phys. B **92** 327
[3] Maulini R, Beck M, Faist J and Gini E 2004 Applied Physics Letters **84** (10) 1659
[4] Wysocki G, Curl R F, Tittel F K, Maulini R, Bulliard J M, Faist J 2005 Appl. Phys. B **81** 769
[5] Pushkarsky M, Weida M, Day T, Arnone D, Pritchett R, Caffey D, Crivello S 2008 SPIE Paper No. 6871-63, Photonics West
[6] Wysocki G, Lewicki R, Curl R F, Tittel F K, Diehl L, Capasso F, Troccoli M, Hofler G, Bour D, Corzine S, Maulini R, Giovannini M, Faist J, 2008 Appl. Phys. B **92** 327
[7] Kosterev A A, Bakhirkin Y A, Curl R F and Tittel F K 2002 Optics Letters **27** 1902
[8] Kosterev A A, Tittel F K, Serebryakov D, Malinovsky A and Morozov I 2005 Rev. Sci. Instr. **76** 043105
[9] Lewicki R, Wysocki G, Kosterev A A and Tittel F K 2007 Optics Express **15** 7357
[10] Bauer C, Willer U, Sharma A K, Pohlkötter A and Schade W, 2008 Laser Applications to Chemical, Security and Environmental Analysis (LACSEA), St. Petersburg
[11] Miklós A, Hess P, and Bozóki Z, 2001 Rev. Sci. Instr. **72**, 1937
[12] Wolfenstein R 1895 Chem. Ber. **28**, 2265
[13] Oxley J C, Smith J L, Shinde K and Moran J, 2005 Propellants, Explosives, Pyrotechnics Vol. **30** 127

Journal of Physics: Conference Series **157** (2009) 012003 doi:10.1088/1742-6596/157/1/012003

Application of quantum cascade lasers and infrared-fibres for the monitoring and control of industrial plasma processes

F Hempel[1,3], V Artyushenko[2], F Weichbrodt[1] and J Röpcke[1]

[1] INP Greifswald, Felix-Hausdorff-Str. 2, 17489 Greifswald, Germany
[2] ART Photonics GmbH, Schwarzschildstrasse 6, 12489 Berlin, Germany

[3]E-Mail: hempel@inp-greifswald.de

Abstract. Quantum cascade lasers (QCL) offer attractive options for the application of mid-infrared absorption spectroscopy for industrial process monitoring and control. In particular the in-situ measurement of reactive plasma compounds can give new insight into the reaction kinetics of industrial plasma processes. For these purposes the compact quantum cascade laser measurement and control system (Q-MACS) has been combined with an infrared-fibre to allow a flexible and completely dust-sealed incoupling of the infrared radiation into the reactor chamber of an industrial plasma surface modification system. The system has been completed with a fast and compact detector and a data acquisition system to monitor concentrations of species of interest in the plasma process online. First tests proofed the potential of this approach.

1. Introduction

Molecular plasmas are valuable tools in plasma enhanced chemical vapour deposition and in etching systems to deposit or remove thin films with advantageous mechanical, electrical or optical properties. By varying the plasma parameters coatings with a wide range of chemical and physical properties can be produced. Therefore and for reasons of enhanced efficiency, increased stability and product quality the direct control of such plasma applications is a goal and challenging subject for plasma technology. The key to an improved understanding and control of plasma chemistry and kinetics in such chemical active discharges is the analysis of the fragmentation of the precursor gas and the monitoring of transient or stable plasma reaction products, in particular the measurement of their ground state concentrations. Information on absolute concentrations of plasma species can be used e.g. for improvements of productivity, reliability and reproducibility.

In plasmas the measurement of absolute molecular species number densities can be done using absorption spectroscopic methods in the mid infrared spectral region. Tuneable infrared diode laser absorption spectroscopy (TDLAS) in the spectral region between 3 and 20 μm is a well known non-invasive technique for the detection and the measurement of concentrations of stable and transient molecular species not only in the gas phase but in gas discharges as well [1,2]. It can also be used to determine neutral gas temperatures and to investigate dissociation processes in a wide range of molecular discharge plasmas. The TDLAS method also allows time resolved measurements.

[3] Author to whom any correspondence should be addressed

The main disadvantage of TDLAS systems is the necessary cryogenic cooling of the lead salt diode lasers and of the detectors, since TDLs operate at cryogenic temperatures below 100 Kelvin. Systems using lead salt diode lasers are typically large in size and require closed cycle refrigerators and/or consumables like liquid nitrogen.

Quantum cascade lasers (QCL) are a new type of mid infrared laser sources, operating at room temperature and having similar spectroscopic properties as TDLs. The output power is sufficient to combine them with thermoelectrically cooled infrared detectors, which permits a decrease of the instrument size and gives the opportunity to design compact liquid nitrogen-free mid-IR spectroscopic systems. "Turn-Key" operation and unattended remote monitoring and control at an improved safety level and at a low level of maintenance could be ensured. These positive features of QCLAS systems can open up new fields of application in research and industry, including studies of gases in atmospheric, environmental and plasma chemistry but also to study and monitor industrial plasma processes.

The applicability of QCLAS has been proven using the quantum cascade laser measurement and control system "Q-MACS" for on-line process monitoring in research and industry. Results of such measurements and features of Q-MACS are given in [3,4,5]. In this paper we present a family of QCLAS measurement systems, which were developed taking especially the requirements of measurements in industrial plasma reactors for boriding of metal parts into account.

2. The Q-MACS Process

The Q-MACS Process is a robust compact and transportable measurement system which was developed and build with the focus on measuring and monitoring borone containing species in industrial plasma process reactors. Figure 1 shows the Q-MACS Process in a test at an industrial plasma reactor; figure 2 illustrates the optical arrangement of the system.

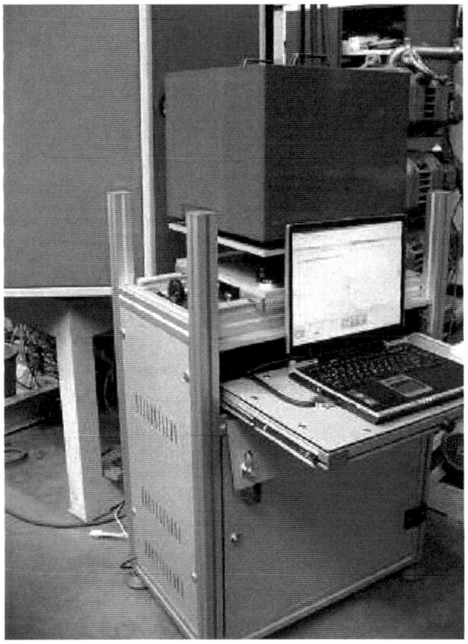

Figure 1. Photo of the Q-MACS Process with industrial plasma reactor in the background

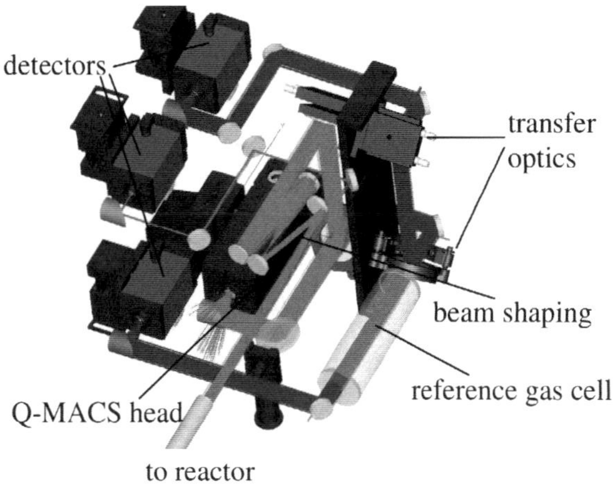

Figure 2. Opto-mechanical scheme and beam propagation simulation of the optical arrangement of the Q-MACS Process

The optical design consists of three independent beam paths to achieve long term stability and low signal-to-noise ratios of the measurement. The infrared laser beam emitted by a QCL, which is mounted in a Q-MACS laser head, is divided by two ZnSe beam splitter plates for this purpose. In addition to the measurement path a pulse normalization path is added to neglect fluctuations of the pulse height of the QCL pulses and a reference locking path is used to assure the stability of the laser frequency by line locking employing a reference gas cell. While the reference locking is good for long term stability of the system, pulse normalization may result in an increase of the device sensitivity of up to an order of magnitude. Beam shaping transfer optics are included in the measurement path to decrease the laser beam diameter to a size which fits to the diameter of the windows of the plasma reactor and which allows to set up a dual-path arrangement through the reactor. In this case a retro reflective mirror is placed on the other side of the plasma reactor.

To achieve a compact size of the optical arrangement a dual platform setup was designed. This is a good compromise between compact dimensions, robust and compact installation, accessibility and easy alignment possibilities.

The measurement signals from the three thermoelectrically cooled infrared detectors are digitized by data acquisition cards. The rack contains the optics, the controlling computer, a thermo chiller and the supplies for all devices. The Q-MACS Process needs an external power source and a retro mirror (in the case of dual path measurements) for operability, only.

Recently, the Q-MACS Process was completed by a long path cell module, holding a 76 m astigmatic Herriot cell, to extend the measurement opportunities of the system to high sensitive trace gas analysis. With this additional board sensitiveness in the ppb range can be achieved.

3. The Q-MACS Process 2

Tests of the Q-MACS Process at the plasma reactor of the industrial partner showed some drawbacks of this system and its approach for the special purpose of measuring species with dense spectra in plasma reactors for industrial boriding of steel parts. Even though the optics were boxed dust contamination of the mirrors of the optical arrangement due to the industrial environment was observed. While the Q-MACS Process rack has a stable and fixed stand besides the plasma reactor the vessel itself moves with its flanges and windows due to the thermal expansion caused by the heating of the wall chambers. The laser beam passes a beam splitter and four times through dirty windows which results after some time of operation in weak measurement signals and insufficient signal to noise ratios due to depositions on the reactor windows. Besides, there is no free choice of the positioning of the Q-MACS Process since the infrared laser beam is leaving the system at a fixed position and angle and has to come back in a defined way (in the case of dual path measurements). This limits the flexibility of positioning the system significantly and blocks the access to the chamber in case of maintenance or part exchange.

Therefore, it was appropriate to design a Q-MACS Process 2, following an adapted approach.

The Q-MACS Process 2 is also designed for monitoring and control of plasma aided industrial surface treatment processes. It combines Q-MACS as the infrared source with an infrared-fibre to allow a flexible and enclosed laser beam handling and remote positioning of the laser source and control unit. The optics are simplified to the greatest possible extent to simplify the alignment and enhance the robustness of the system. Therefore no normalization or reference locking path is used. The QCL emission is directly coupled into the infrared-fibre by a compact and sealed set of optics. The incoupling of the laser beam into the plasma reactor and the detection of the signal is done by compact and light weight modules which are directly mounted to the reactor flanges. Since these enclosed modules are dust-sealed as well, the complete optical arrangement does not have any open optical paths outside the reactor. This is in combination with the flexible positioning of the system rack an important advantage and improvement of the Q-MACS Process 2, using infrared-fibres. Figure 3 shows the scheme of the experimental arrangement and figure 4 a photo of the device at an industrial plasma reactor.

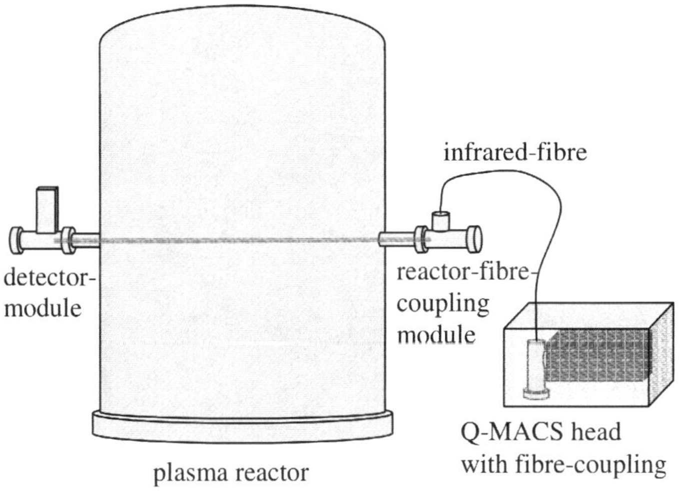

Figure 3. Experimental arrangement of the Q-MACS Process 2 with IR-fibre coupling to access the industrial plasma reactor

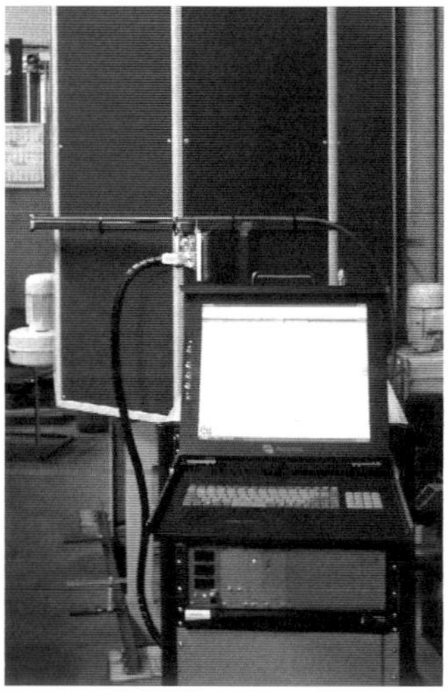

Figure 4. Photo of the Q-MACS Process 2 with detector module and industrial plasma reactor

The housing of the Q-MACS laser head with its fibre coupling module (figure 5) is integrated in an industrial standard 19-inch-rack (figure 6). The rack contains the Q-MACS supply, the laser head and a computer system with fast data acquisition boards. The rack is connected to the reactor-fibre-coupling module by a commercial infrared-fibre. This coupling module is a dust-sealed set of optics to parallelize the infrared beam and to guide it through the plasma chamber. On the other side of the reactor the detector module is located. It consists of dust-sealed optics which focuses the laser beam on the detector. Since this detector allows temporal resolutions in the nanosecond-range it is possible to measure in intrapulse mode as well as in the interpulse mode. Temporal resolutions of the measured concentration development down to the microsecond-range can be achieved. The detector signal is digitized by fast data acquisition boards in the computer system.

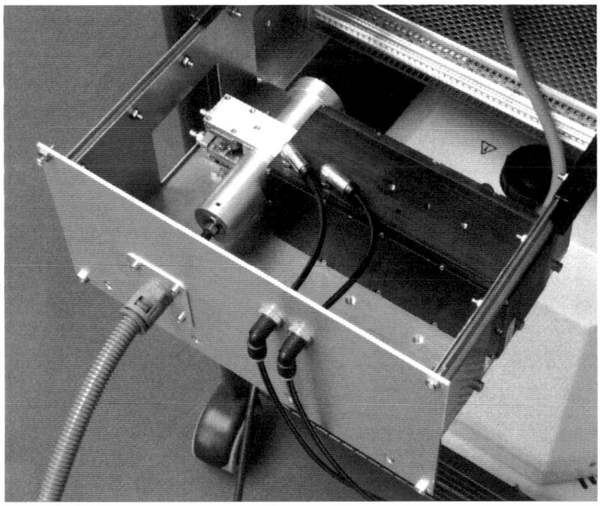

Figure 5. Photo of the Q-MACS laser head with the fibre coupling module of the Q-MACS Process 2

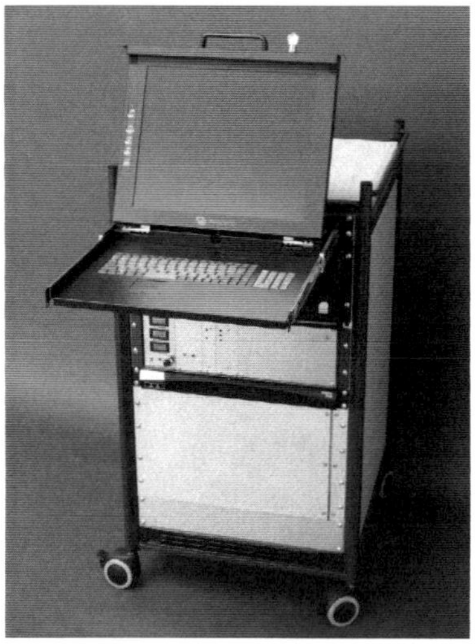

Figure 6. Photo of the Q-MACS Process 2 with integrated fibre coupling

4. Examples of applications

4.1. Concentration monitoring of gases in industrial plasma processes

QCLs emit radiation of narrow bandwidth (0.01 cm^{-1}) leading to a high spectral resolution over a typical tuning range of several wavenumbers. Therefore with QCLs single vibration-rotational absorption lines can be measured. In this case Beer-Lambert's law can be used to determine the molecular concentrations, assuming homogeneity over the absorption path and neglecting absorption saturation effects:

$$I(v) = I_0(v)\exp[-\sigma(T,p,v)nl]$$

$I(v)$ is the transmitted intensity, $I_0(v)$ is the irradiated intensity, $\sigma(T,p,v)$ is the temperature-and pressure dependent absorption coefficient of a specific absorption line of the frequency v with a typical dimension of a cross section per molecule [cm^2 $molecule^{-1}$], n is the total number of molecules and l is the absorption length. The absorption coefficient can be written as the product of a line strength S of a specific line and of the normalized line profile function f:

$$\sigma(v) = S\ f(v - v_0)$$

with

$$\int_{line} f(v - v_0)dv = 1\,.$$

While the influence of the line profile f can be calculated, the value of the line strength S has to be taken from literature [6] or to be determined.

Several molecular gases, which are used in industrial processes, show absorption spectra which can not be spectrally resolved at room temperature caused by a variety of overlapping absorption lines, e.g. in case of different isotopomers of high abundance. For example BCl_3 combines ^{11}B (80.1 %) and ^{10}B (19.9 %) with ^{35}Cl (75.8 %) and ^{37}Cl (24.2%). The molecules formed by the different combinations of these isotopes have marginally differing spectra overlapping each other in the spectral range between 920 cm^{-1} and 1010 cm^{-1} (figure 7). Despite the high spectral resolution the resulting spectrum can not be resolved; it is a dense, multi-peak spectrum without any zero-absorption parts in between (figure 8).

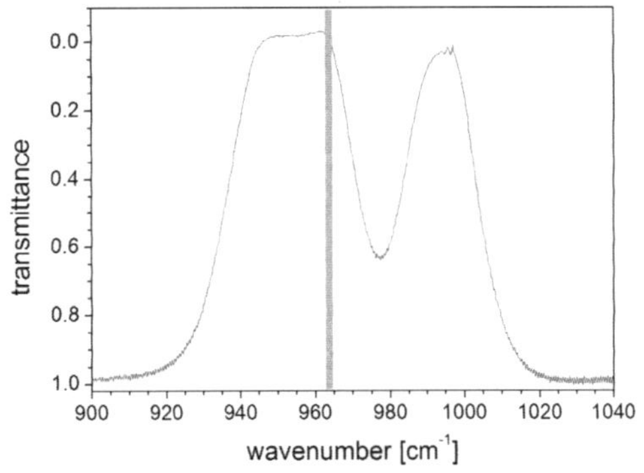

Figure 7. FTIR spectra of BCl_3 in the spectral range 900 cm^{-1} to 1040 cm^{-1} with measurement position (grey)

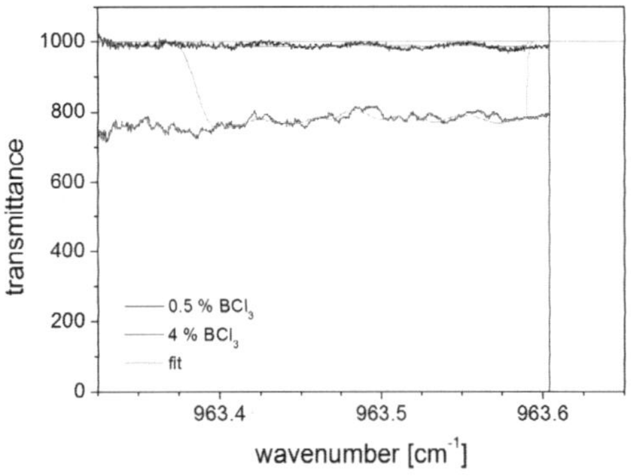

Figure 8. QCLAS spectrum (black, blue) of BCl_3 at 963.5 cm^{-1} with multi peak fit (red)

For such complex spectra normally no line strength data are available in literature. Therefore, special routines have to be used to determine the line strength data experimentally for quantitative measurements of these species.. For this purpose in the present case the spectrum of BCl_3 was recorded and analysed by means of FTIR spectroscopy first (figure 7 shows the part with the main absorption band of BCl_3). In the most intense part of the band HITRAN-format files containing line positions and strengths [6] were estimated from highly resolved spectra, recorded between 963 cm^{-1} and 965 cm^{-1} using a tunable diode laser (TDL) absorption spectrometer [7]. Particular attention was given to the calibration of the wavenumber scale and the correct gas mixture and pressure. More details on handling complex spectra can be found in [8,9]. Tests of the deduced HITRAN-format files showed a good agreement of measured and fitted spectra regarding the position and absorption strength of mean spectral features of BCl_3 in this range. Figure 8 shows results of tests of a part of this file using 292 line entries to fit the measured data taken with a Q-MACS Process spectrometer.

Stancu et al [5] used another approach for non-resolvable spectra in the mid-infrared region. Instead of determining line strengths they use the absorption coefficient $\sigma(T,p,v)$ for a spectral position at the temperature and pressure of interest for the measurement. The absorption coefficient has to be determined before experimentally. Since industrial plasma processes usually run at given pressures and temperatures this approach is possible even although it limits the flexibility for changes of process parameters. Assuming constant values of the absorption coefficient σ, of the absorption length l and of the intensity I_0 irradiated from the laser it is sufficient to measure the transmitted intensity I only for getting absolute concentration values.

4.2. Results with Q-MACS Process

The data acquisition unit of Q-MACS Process was designed for the requirements of the TDLWintel software package [10,11]. This program can be used to control QCLs, e.g. using the Q-MAC system, to analyze the recorded spectra in real time and to display results of measured concentrations. By using this rapid scan software with real-time line shape fitting routines a time resolution of up to 1 µs can be provided, allowing the study of kinetic processes of infrared active compounds in plasmas or gases. To increase the signal to noise ratio the measurement can be averaged for several seconds, being a sufficient temporal resolution for most of the industrial plasma process monitoring applications. With TDLWintel, using the prerecorded HITRAN-file it was possible to monitor the development of the concentration of the precursor gas BCl_3 in an industrial pulsed dc plasma reactor while driving process plasmas to deposit borone based hard coatings. The measurement results of the BCl_3 percentage in the chamber are not only displayed but can be transferred via hardware (serial) or software (network TCP/IP) interfaces to other systems, e.g. for the use in automated process monitoring and controlling. This was tested using the Q-MACS Process.

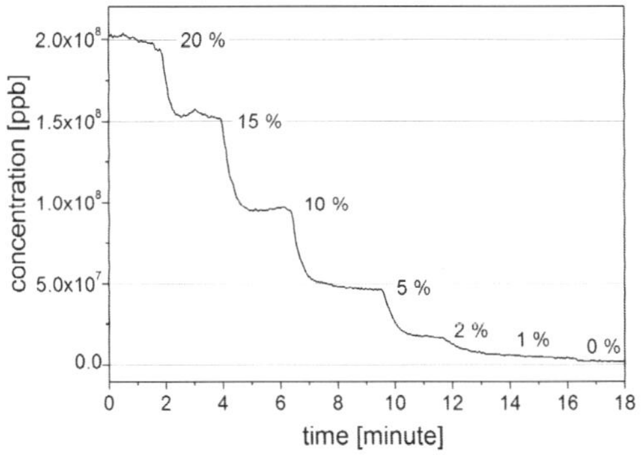

Figure 9. Measurement of the BCl_3 concentration in the gas inlet with Q-MACS Process

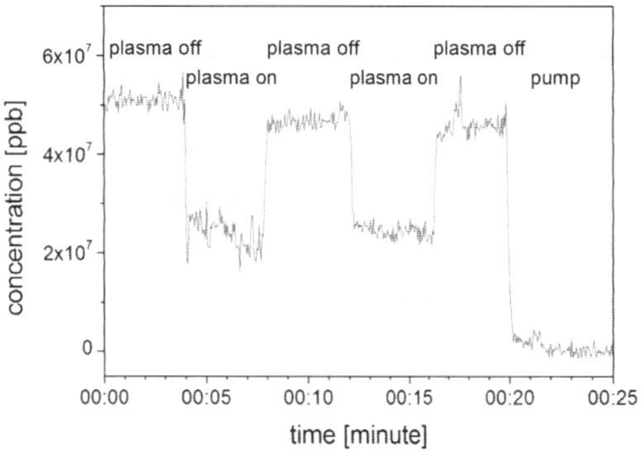

Figure 10. Comparison of the BCl_3 concentration for plasma on and off phase in a microwave discharge reactor

First test measurements of well known BCl_3 concentration in the gas inlet were performed with the Q-MACS Process without starting a plasma. Figure 9 shows the measured concentration steps of decreasing BCl_3 content in a mixture with hydrogen and argon. Using TDLWintel it was possible to reproduce the admixed ratios within sufficient error limits allowing to monitor the concentration change of BCl_3 permanently and over a wide range of concentrations while driving plasma processes.

In figure 10 the temporal development of the BCl_3 concentration during a test in a microwave discharge chamber is shown. In such mixtures with Argon the ignition of the plasma resulted in an increase of the degree of dissociation of more than 50 %. The BCl_3 concentration signal could be followed after switching off the plasma, the gas inlet and the starting to pump for about 2 minutes.

Using a QCL the scan through an infrared spectrum may be commonly achieved in two different methods. In the *inter* pulse mode, used by TDLWintel, a bias DC ramp is applied to a series of short laser pulses of a few tens of nanoseconds. Another option is the *intra* pulse mode, i.e. the scanning in single, longer pulses acquiring an entire spectrum in this pulse. Since this scan is performed in tens up to a few hundred nanoseconds with duty cycles of the QCL of several percent time resolutions in the microsecond range have become possible for quantitative in-situ measurements of molecular concentrations in plasmas with QCLs as well.

4.3. Results with Q-MACS Process 2

The Q-MACS Process 2 system has been designed for improved adaptation to industrial applications. The system allows the same measurements as described in chapter 4.2. Concentration values of various molecular species can be measured with temporal resolutions from μs to s in the intrapulse mode as well as in the interpulse mode. Figure 11 shows the change of the BCl_3 concentration in the gas flow during such test measurements. In comparison to the calculated BCl_3 concentration determined by fitting the complex spectra using TDLWintel (red) the change of the recorded signal (blue) is displayed as well. The similarity of both results is a good proof for the approach to measure the transmitted signal only for monitoring species concentration values.

Figure 12 shows an example of a plasma process monitoring with various contents of BCl_3 in the gas mixture and for various plasma duty cycles. Even though fluctuations caused by long-term changes of the irradiated laser intensity were observed, these results proofed that it is possible to monitor industrial plasma processes using the Q-MACS Process 2. For further improvements of the signal stability, of the signal to noise ratio and of the measurement sensitivity it is planned to implement a simplified signal normalization path into the optical arrangement.

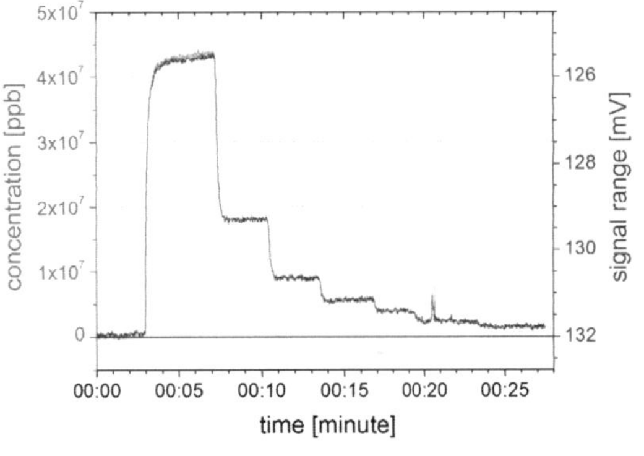

Figure 11. Comparison of the BCl_3 concentration (red) in the gas inlet with the signal (blue)

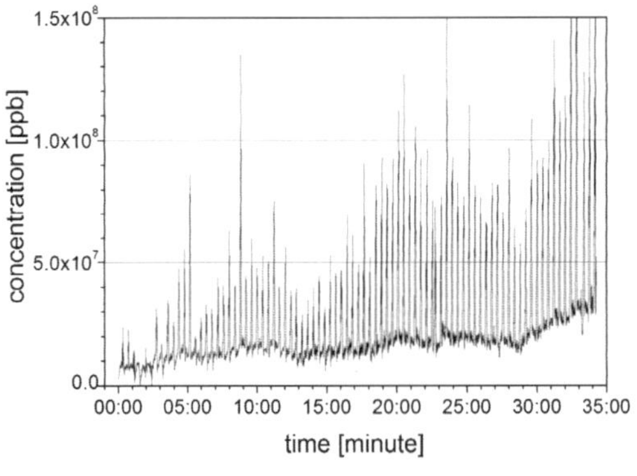

Figure 12. Change of the BCl_3 signal in an industrial plasma process

5. Acknowledgements
This work was partly supported by the Deutsche Forschungsgemeinschaft, Transferbereich 36, by the Bundesministerium für Bildung und Forschung, FKZ 13N7451/8 and by the Bundesministerium für Wirtschaft und Arbeit, FKZ KF 0086503FK6. The authors give sincere thanks to Charles Harward from Nottoway Scientific Consulting Cooperation, Nottoway USA, for his valuable support in handling dense absorption spectra. In particular, the authors are indebted to S. Strämke (ELTRO GmbH, Baesweiler, Germany) for the successful cooperation over several years. K.-D. Weltmann is gratefully acknowledged for his continuous encouragement and general support making the present work possible.

References
[1] Röpcke J, Mechold L, Käning M, Fan W Y and Davies P B 1999 *Plasma Chem. Plasma Process.* **19** 395
[2] Hempel F, Davies P B, Loffhagen D, Mechold L and Röpcke J 2003 *Plasma Sources Sci. Technol.* **12** S98
[3] Hempel F, Glitsch S, Röpcke J, Saß S and Zimmermann H 2005 *Plasma Polymers and Related Materials* ed M Mutlu 142
[4] Röpcke J, Glitsch S, Hempel F, Lang N, Macherius U, Saß S, Stancu G D, Weichbrodt F, Weltmann K-D and Welzel S 2006 *VDI-Berichte* **Nr 1959** 279
[5] Stancu G D, Lang N, Röpcke J, Reinicke M, Steinbach A and Wege S 2007 *Chem. Vap. Deposition* **13** 351
[6] Rothman L S, Jacquemart D, Barbe A, Benner D C, Birk M, Brown L R, Carleer M R, Chackerian C, Chance K, Coudert L H, Dana V, Devi V M, Flaud J M, Gamache R R, Goldman A, Hartmann J M, Jucks K W, Maki A G, Mandin J Y, Massie S T, Orphal J, Perrin A, Rinsland C P, Smith M A H, Tennyson J, Tolchenov R N, Toth R A, Vander Auwera J, Varanasi P, Wagner G 2005 *Journal of Quantitative Spectroscopy & Radiative Transfer* **96 (2)**: 139
[7] Röpcke J, Mechold L, Käning M, Anders J, Wienhold F G, Nelson D D and Zahniser M 2000 *Rev. Sci. Instrum.* **71** 3706
[8] Harward C N, Baren R E and Parrish M E 2004 *Spectrochem. Acta Part A* **60** 3421
[9] Harward C N, Thweatt W D, Baren R E and Parrish M E 2006 *Spectrochem. Acta Part A* **63** 970
[10] Zahniser M S, Nelson D D and Kolb C E 2002 in Applied Combustion Diagnostics, Kohse-Hoinghaus K and Jeffries J (eds), Tallor and Francis, New York 648
[11] Nelson D D, Shorter J H, McManus J B and Zahniser M S 2002 *Appl. Phys. B* **75** 343

Journal of Physics: Conference Series **157** (2009) 012004 doi:10.1088/1742-6596/157/1/012004

On the Destruction of Volatile Organic Compounds using a Dielectric Pellet Bed Reactor

M Hübner and J Röpcke

INP Greifswald, Felix-Hausdorff-Str. 2, 17489 Greifswald, Germany

E-mail: huebner@inp-greifswald.de

Abstract. Volatile organic compounds (VOC) are known to be dangerous for human health. If they are part of a precursor gas mixture undesired effects in e.g. industrial processes may occur. Therefore, the destruction of unwanted VOCs in a gas stream is an important aim. For that purpose a five stage planar dielectric barrier discharge reactor working at atmospheric pressure has been developed and tested. Each stage consists of two parallel stainless steel meshes serving as electrodes. The space between them is filled with dielectric beads. The ac voltage applied to the electrodes is in the kV and kHz range and several gas flow rates are possible. The gases of interest have been measured using a Fourier Transform Infrared Spectrometer (FTIR) combined with a 32.5m long path cell. First result showed, that the VOC ethylene (C_2H_4) can be removed.

1. Introduction

Volatile Organic Compounds (VOCs) are chemical substances which are easy to vaporise and therefore, when exhausted, are present in the gas phase at room temperature and ambient pressure. Although no unique and precise definition exists these substances can be roughly classified into four groups based on their boiling points [1]:

Table 1. A rough classification of VOCs.

Name	Boiling Point [°C]
Very Volatile Organic Compounds (VVOC)	<0 - 50...100
Volatile Organic Compounds (VOC)	50...100 - 240...260
Semi Volatile Organic Compound (SVOC)	240...260 - 380...400
Organic compound associated with particulate matter or particulate organic matter (POM)	380

As one can see from table 1 a lot of these substances are vaporised at room temperature and therefore, if present, easy to breath.

Nowadays two main sources can be distinguished: natural sources such as wetlands, ruminants or rice agriculture and anthropogenic sources such as paint thinners, petroleum fuels or solvents. Whereas natural sources are considered to be unproblematic the man-made VOCs emitted e.g. from chemical industries are considered to be critical because some of these VOCs are suspected or known to cause cancer or to damage the central nervous system of humans. Therefore anthropogenic emission of VOCs into the atmosphere or into ground water can lead to certain health problems for humans. For example, the VOC toluene is suspected to harm the embryo. Accumulation of VOC's in the atmosphere or in the ground water has to be avoided because of these facts. Therefore waste gas exhaust from industry has to be cleaned.

Plasma processing for air pollution control has a high potential for industrial applications. Based on a new approach by Whitehead and co-workers [2] a new five stage packed-bed dielectric barrier discharge reactor working at atmospheric pressure has been designed and tested. Harling et. al. found that such an arrangement is appropriate for efficient destruction of VOC's inside a gas stream without NO_x production and that three stages working in series are more efficient compared to a parallel arrangement [2]. This synergistic effect will be investigated in more detail using infrared absorption experiments. Fourier Transform Infrared Spectrometry (FTIR) is used for overview spectra, and for higher sensitive species concentration measurements Tuneable Diode Laser Absorption Spectroscopy (TDLAS) and Quantum Cascade Laser Absorption Spectroscopy (QCLAS) will be applied. Within the last years both techniques have been successfully established for studying molecular phenomena in plasmas and gases, in particular in kinetic and chemical processes [3, 4, 5].

In the present work the VOC ethylene (C_2H_4) was used as a pollutant. Ethylene is gaseous at ambient conditions with a boiling point of -103°C and a melting point of -169°C [6]. When humans are exposed to low concentrations of ethylene narcotic effects like headache, vertigo and loss of coordination may occur.

This paper describes a new five stage reactor and presents first results about the ethylene destruction in an air stream.

2. Experimental Setup

The new five stage packed-bed plasma reactor is made of acryl glass as shown in figure 1. The dimensions over all are 390 mm x 130mm x 130mm. The main units are the five stages. Each stage is made of two stainless steel meshes which serves as electrodes. The space between them is filled with glass beads with a diameter d = 6mm. The dimension of one electrode is 109mm x 109mm x 1.1mm. The space between two electrodes is approximately 15mm. The distance between the stages is about 45mm, figure 2. The top cover is removable. The gas supply can be realized by a tube connection at the front walls (left and right side in figure 1). Gas samples can be taken via tube connections between each stage. Additionally, IR transparent windows (KBr) are placed between each stage for absorption spectroscopy (AS) purposes.

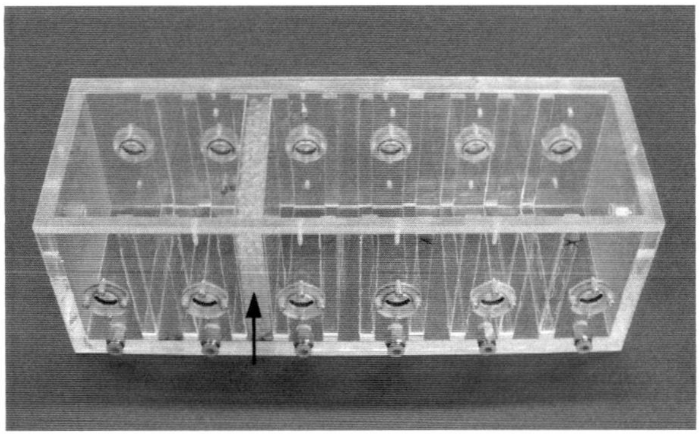

Figure 1. Photograph of the new five-stage reactor. Only one stage is filled here, indicated by the arrow.

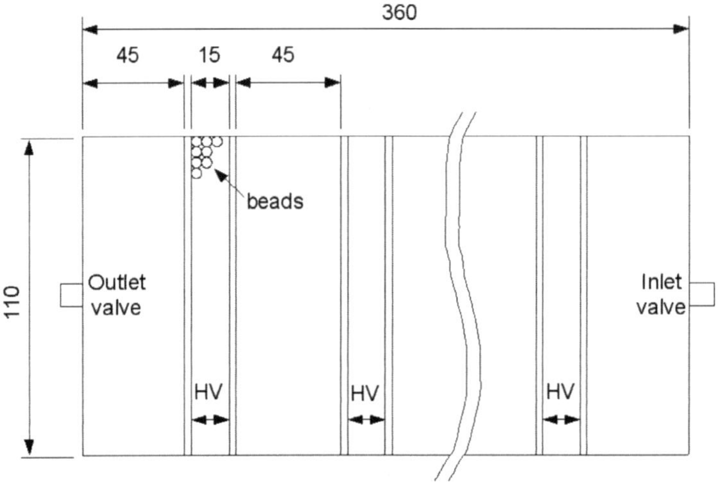

Figure 2. Principal sketch of the new five-stage reactor. All dimensions are in millimetre.

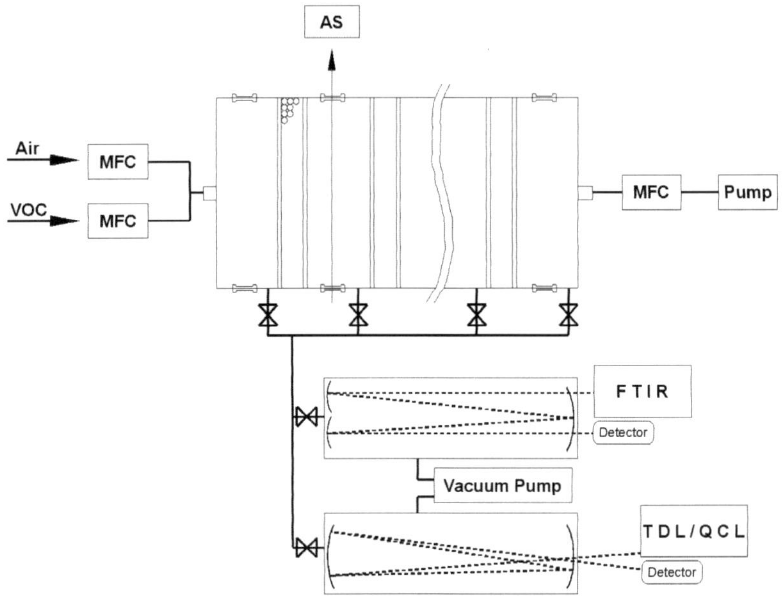

Figure 3. Schematic of the whole experimental set-up. It is planned to use TDLAS and QCLAS techniques, as well as FTIR, in future investigations.

Each stage was separately driven by a NeonPro (NP-10000-30) ac power supply with a voltage of U_{HV} = 6500V ac and a frequency of f_{HV} = 20kHz. The power was measured with a commercial power meter connected before the NeonPro power supply. The power measured in this way is the system power for one stage. It was measured to be P_{system} = 8W per stage.

The gas flow guided into the inlet valve of the reactor consisted of 1slm air and 1sccm C_2H_4 (99.95 vol. % purity) which is equivalent to an ethylene concentration of about 1‰. It was controlled with mass flow controller units (MKS), premixed and then guided into the reactor. The outlet gas stream was guided into a 32.5m long path cell (White cell) and investigated using a Fourier Transform

Journal of Physics: Conference Series **157** (2009) 012004 doi:10.1088/1742-6596/157/1/012004

Infrared Spectrometer (Bruker IFS66/v). A vacuum pump was used to stabilize the pressure inside the long path cell at 50mbarr.

3. Results

Two FTIR spectra of the gas at the outlet of the reactor are shown in figure 4. One is from C_2H_4 after gas flow stabilisation and the second was recorded 10min after the plasma was ignited. As one can see in figure 4, ethylene has been almost completely removed.

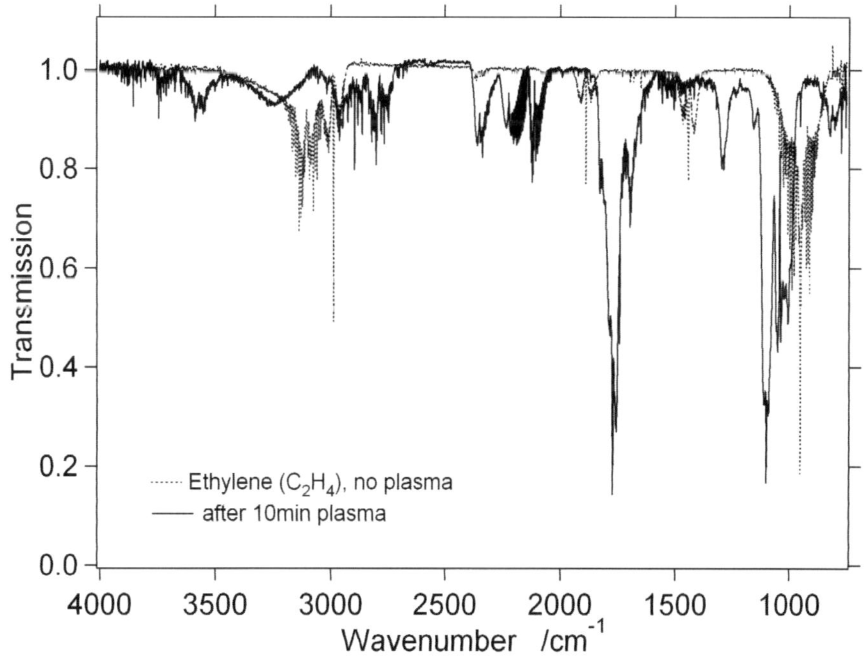

Figure 4. FTIR spectra of (i) the gas stream with ethylene without plasma and (ii) of the outlet gas flow treated by plasma in the reactor. Φ= 1slm air + 1sccm C_2H_4, P = 8W.

The identification of the species produced in the plasma was done using the HITRAN [7] database and the software IGOR was used to simulate the spectra. The parameters for the simulation were: pressure p = 50mbar, absorption length l = 3250cm, temperature T = 296 K and line width $l_w = 1cm^{-1}$.
Typical products of the plasma treatment were found to be CO, CO_2, N_2O, O_3, H_2CO and H_2CO_2. In figures 5-7 detailed parts of measured FTIR spectra are shown and compared with simulated spectra to identify the products of the plasma treatment and to estimate their concentrations.

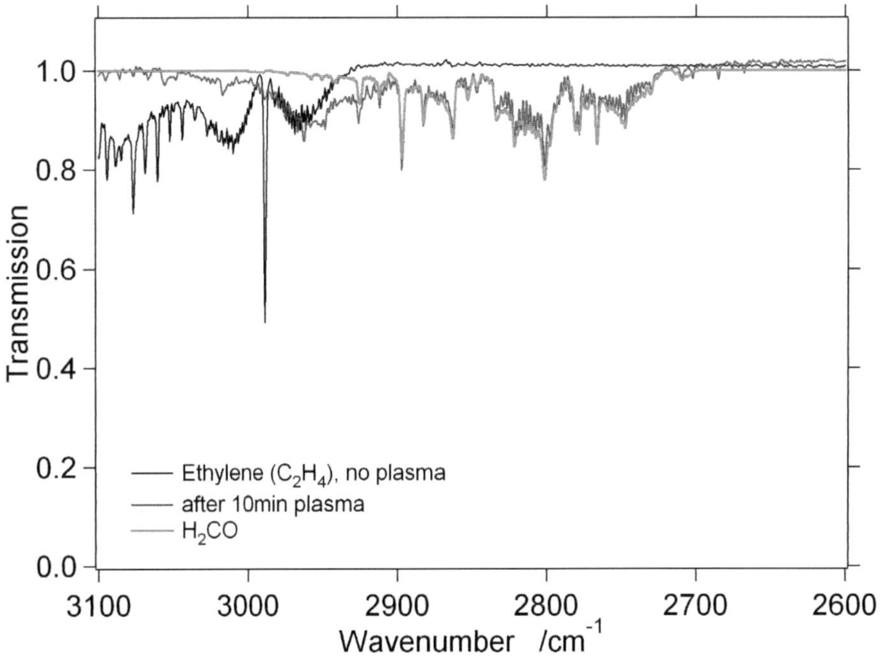

Figure 5. Identification of H_2CO. Species concentration used for spectrum simulation: 280ppm H_2CO.

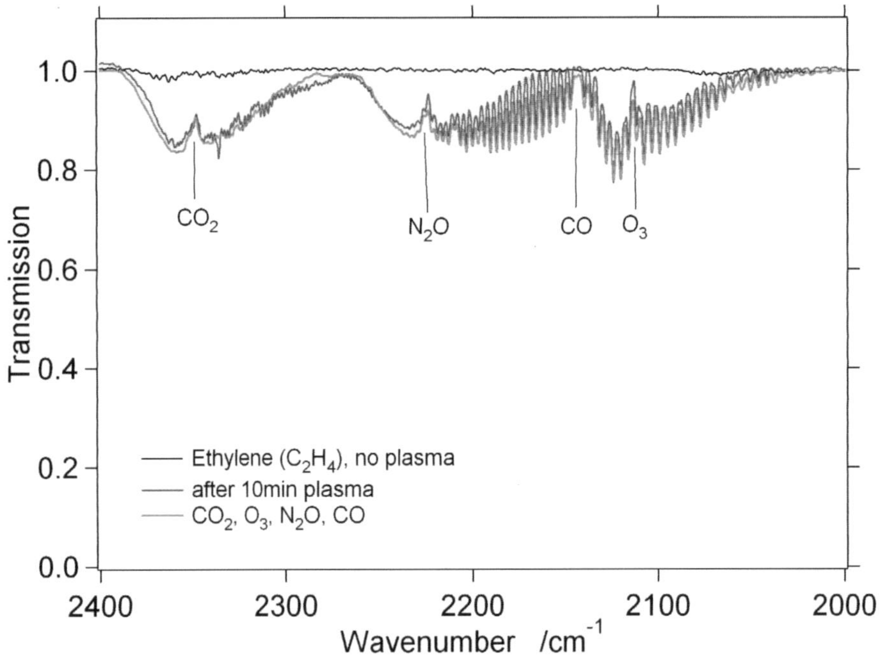

Figure 6. Identification of CO_2, N_2O, CO and O_3. Species concentration used for spectrum simulation: 170ppm CO_2, 900ppm O_3, 100ppm N_2O and 0.4% CO.

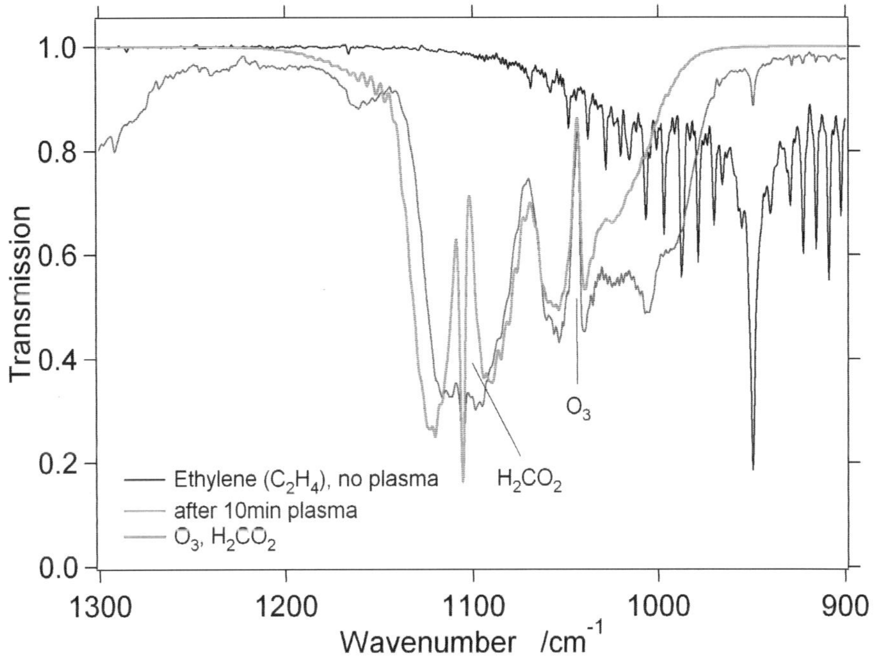

Figure 7. Identification of O_3 and H_2CO_2. Species concentration used for spectrum simulation: 900ppm O_3, 0.15% H_2CO_2.

4. Conclusion / Outlook

This work has shown that the new packed-bed plasma reactor works in principle and can be used to remove ethylene from a gas stream. Typical products of the plasma treatment were found to be CO, CO_2, N_2O, O_3, H_2CO and H_2CO_2.

Further work will focus on a detailed investigation about the ethylene destruction. One experiment will use a different number of stages to investigate the synergistic effect reported by [2, 8]. For that purpose the new reactor permits an easy variation in the number of stages as well as different distances between them. This allows one to realize different residence times of the chemicals between consecutive stages. Whitehead and co-workers assume that one stage produces for example O_3 which destroys the pollutant chemically on the way through the reactor.

Acknowledgement

The authors are gratefully acknowledging S. Saß und K.D. Schulz for technical support. K.-D. Weltmann is gratefully acknowledged for his continuous encouragement and general support.

References

[1] http://de.wikipedia.org/wiki/Fl%C3%BCchtige_organische_Verbindungen 2008

[2] Harling A M, Glover D J, Whitehead J C and Zhang K 2008 Novel Method for enhancing the destruction of environmental pollutants by the combination of multiple plasma discharges *Env. Sci. Tech.* **42** 4546

[3] Welzel S, Gatilova L, Röpcke J and Rousseau A 2007 *Plasma Sources Sci. Technol.* **16** 822–31

[4] Stancu G D, Lang N, Röpcke J, Reinicke M, Steinbach A and Wege S 2007 *Chem. Vap. Deposition* **13** 351–60

[5] Röpcke J, Lombardi G, Rousseau A and Davies P B 2006 *Plasma Sources Sci. Technol.* 148–68.

[6] Air Liquide http://www.airliquide.de/loesungen/produkte/gase/gasekatalog/stoffe/ethen.html#ethen-eigenschaften

[7] HITRAN database http://cfa-www.harvard.edu/HITRAN/

[8] Zhang H 2007 *Non-Thermal Plasma Destruction of Environmental Pollutans* (Manchester: University Library)

[9] Wagner H E, Brandenburg R, Kozlov K V, Morozov A M and Michel P 2005 *Contrib. Plasma Phys.* **45** 338–47

Journal of Physics: Conference Series **157** (2009) 012005 doi:10.1088/1742-6596/157/1/012005

Investigation of atmospheric pressure nitrogen plasmas by cavity ring down spectroscopy

F Kaddouri[1,2], G D Stancu[1], D A Lacoste[1], and C O Laux[1]
[1]Laboratoire EM2C, CNRS UPR288, Ecole Centrale Paris, Châtenay-Malabry, France

Email: farah@em2c.ecp.fr

Abstract. We have applied Cavity Ring-Down Spectroscopy (CRDS) to investigate Nanosecond Repetitively Pulsed Discharge (NRPD) in nitrogen and air at atmospheric pressure in a pin-to-pin electrode configuration. We have measured the concentration of $N_2(A)$ by probing absorption transitions in the near Infrared at 770 nm. We report on preliminary results about the measurement of the $N_2(A)$ concentration and its evolution under different discharge regimes. We present here a comparison for an air plasma between the time evolution of $N_2(A)$ measured by CRDS and the temporal evolution of atomic oxygen measured by two-photon absorption laser-induced fluorescence (TALIF). This study will allow us to compare our experimental results to those given by a kinetic model and help provide a better understanding of the plasma physicochemical processes.

1. Introduction

Environmental issues are today one of the major societal concerns. As a consequence, new engine development is becoming increasingly subject to strict pollution regulations and to a necessity to reduce fuel consumption. Getting the best compromise between the decrease in toxic emissions and the increase in engine power output without raising production costs, becomes a major challenge. One of the possibilities allowing the reduction of these pollutants consists in using lean premixed combustion systems. However, these systems lead to other problems, such as the occurrence of flame extinction and combustion instabilities, especially destructive to the combustion chamber. To resolve these problems, it was shown [1] that one can enhance lean combustion by a local addition of energy that allows a stabilization of the flame by modifying the thermal and reactive mechanisms. The use of high voltage nanosecond repetitively pulsed discharges (NRPD) represents a promising method in answer to this need. The objective of our study is to determine the mechanism that leads to flame stabilization. Following the results obtained using the kinetic model [2] it was predicted that atomic oxygen, which is a key species in the combustion process, is produced through a two-step mechanism. According to this mechanism, electronically excited molecular nitrogen formed via electron-impact excitation, dissociates molecular oxygen, thus generating atomic oxygen which can oxidize the fuel. To validate the kinetic model, the densities of ground state atomic oxygen [3] as well as the excited electronic state $N_2(A)$ have been measured using two advanced techniques: Two-Photon Laser Induced Fluorescence (TALIF) at 225 nm for detection of atomic oxygen in the ground state, and pulsed Cavity Ring-Down Spectroscopy (CRDS) for the detection of $N_2(A)$ species. CRDS is a very sensitive diagnostic tool for low species concentration measurements. The technique is based on measurement of the decay intensity of a laser beam inside a cavity comprised of two high-reflectivity mirrors. We first measure the decay intensity in an empty cavity and then when the cavity contains the absorbing sample. The difference of these decays gives the coefficient of absorption and if the line strength is known we can calculate the absorbing species concentration. A detailed description of the CRDS technique is given in Ref. [4].

2. Experimental setup

The discharges were generated in a pin-to-pin electrode configuration by high voltage pulses of 5 to 8 kV in amplitude, 10 ns is duration, and applied at a repetition frequency of 10 kHz. The experiments were conducted in atmospheric pressure air and nitrogen flows preheated to 1000 K, with an axial flowrate up to 30 l/min. The plasma discharge was produced by applying nanosecond high-voltage pulses between two metallic electrodes. The distance between these electrodes in pin-to-pin configuration was 3 mm.

The cavity consists of two concave high reflectivity mirrors (R > 99.97%) from Los Gatos Research. The distance between the cavity mirrors is 67 cm. The cavity is fed by a tunable pulsed dye laser (Continuum ND 6000) pumped with a Continuum Precision PL 8010 Nd:YAG laser, with repetition frequency of 10 Hz. The output beam passes first through a mode matching optics system and then enters the CRDS cavity. The light collected after the second cavity mirror is detected with a photomultiplier tube (PMT Hamamatsu H9035-03) and recorded onto the oscilloscope. The collected light is filtered by two bandpass interference filters (CVI) with peak transmittance at 770 nm (CVI-F10-770, FWHM 10 nm). Synchronization of the laser pulse with the discharge pulse is achieved with a Berkeley Nucleonics (BNC) 555 four-channel pulse delay generator, which triggers all devices. The experimental set up is depicted in figure 1.

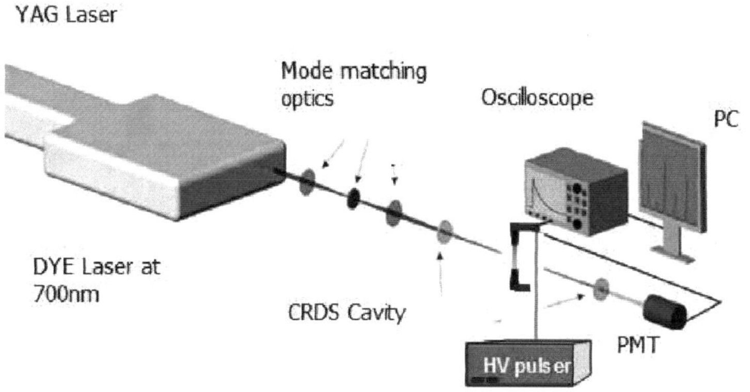

Figure 1. CRDS Experimental Set Up
PMT: photomultiplier tube, HV: high voltage pulse generator

3. Results and discussion

Figures 2 and 3 show typical CRDS signal intensities in linear and logarithmic scale, measured in the nitrogen plasma preheated at 1000 K.

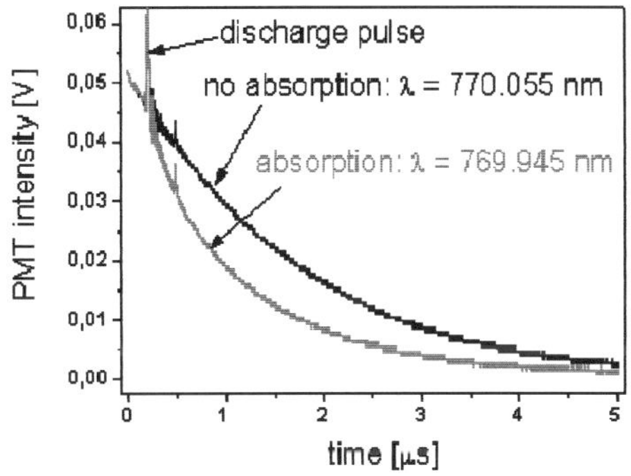

Figure 2. Sample ring-down signals in preheated nitrogen (linear scale)

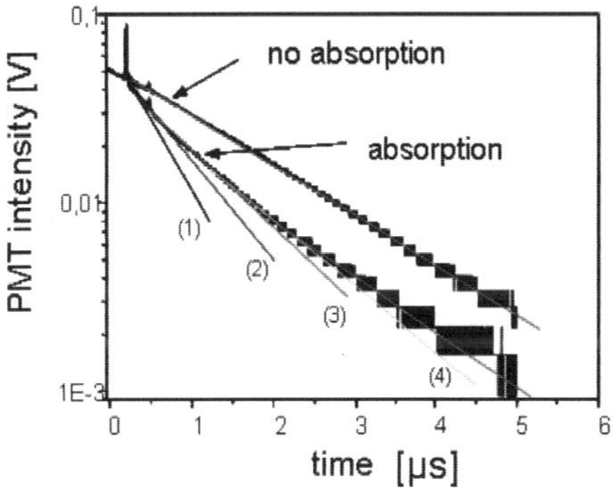

Figure 3. Sample ring-down signals in preheated nitrogen (logarithmic scale)

The black curves show the variation of the intensity recorded by the PMT when the cavity is subjected to 770.055-nm radiation; which corresponds to an off-resonance wavelength for the N_2 B-A transition. The red curves are obtained on an absorption resonance at 769.945 nm. The peak absorption at 769.945 nm is the convolution of $Q_1(22)$ and $Q_3(16)$ transitions of the band: $v' = 2 \leftarrow v'' = 0$. In logarithmic scale, we notice several slopes in the absorption plot which means that the absorption is not constant with time (thus the decay is not a single exponential function of time). The time dependent absorption coefficient is given by the equation (1):

$$k(\nu,t) = \frac{L}{c\,l_{abs}} \frac{d}{dt}\left(\ln \frac{I_1}{I_3} \right) \tag{1}$$

where c is the speed of light, l_{abs} is the absorption length, L is the cavity length, I_1 is the ring-down intensity off-resonance, and I_3 is the ring-down intensity on-resonance.
Using line strengths obtained with Specair [5] the time evolution of $N_2(A)$ concentration could be measured.

Next, we examine how different experimental parameters can influence the $N_2(A)$ concentration. We started first by measuring the variation of $N_2(A)$ concentration with the applied voltage. Figure 4 shows the results averaged over one microsecond immediately after the discharge pulse. . In Figure 4, the $N_2(A)$ concentration increases linearly when the voltage increases from 6.2 to 7.7 kV. Over this voltage range, note that the plasma produced by the discharge varies from a diffuse regime to a filamentary regime. Over this voltage range, the number of $N_2(A)$ molecules increases from $2\text{x}10^{13}$ to $1.4\text{x}10^{14}$ cm^{-3}.

The $N_2(A)$ spatial distribution, shown in Figure 5, has a nearly Gaussian profile, of Full Width at Half Maximum equal to 0.9 mm. In the following analysis, we therefore assumed that the $N_2(A)$ species are approximately uniform in a cylinder of about 1 mm in diameter.

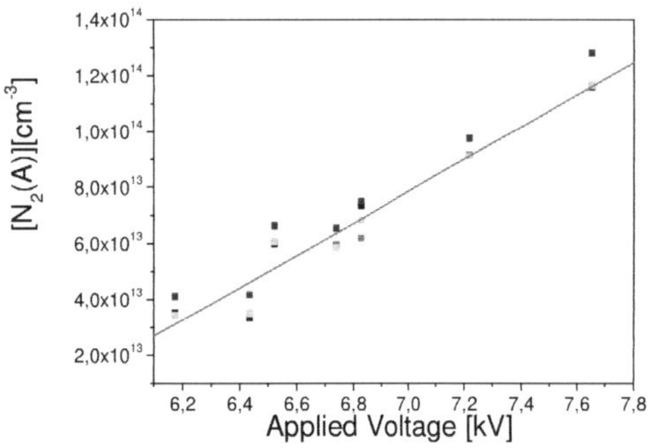

Figure 4. Variation of $N_2(A)$ with applied voltage

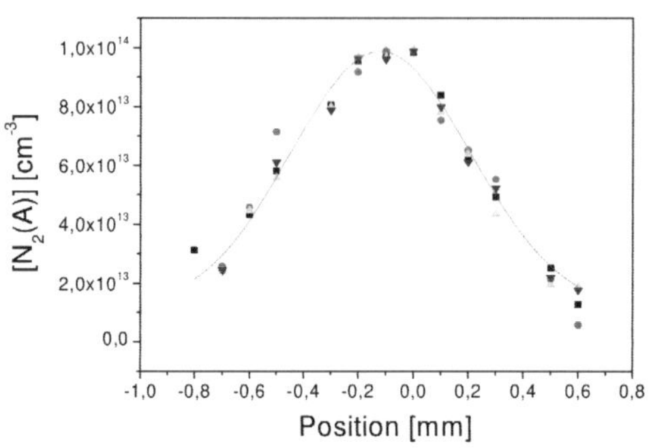

Figure 5. Spatial $N_2(A)$ distribution

4. Conclusion

CRDS was successfully applied to measure the concentrations of key species in nanosecond repetitively pulsed discharges in nitrogen and air at atmospheric pressure with high spatial (sub millimeter) and temporal (sub microsecond) resolution. CRDS has proven to be a sensitive technique for $N_2(A)$ detection.

In nitrogen, the $N_2(A)$ density is found to increase linearly with the applied voltage. In air, the characteristic time of $N_2(A)$ concentration decay coincides with characteristic time for atomic oxygen formation about 100 ns after the discharge event, suggesting that atomic oxygen production is the result of a two-step mechanism as predicted by the kinetic model. This is a partial validation of the two-step mechanism. For a complete validation, a calibration of the TALIF signal using xenon as a reference for absolute atomic oxygen densities is in progress.

Acknowledgments
This work was supported by the ANR IPER project and by a Chaire d'Excellence from Ministère de l'Enseignement Supérieur et de la Recherche.

References

[1] Grandin G A, Grisch F, Vervisch P 2008 Journée des thèses Onera physique, Combustion assistée par plasma en milieux diphasique application au moteur aéronautique.
[2] Pai D, PhD thesis 2008 Étude des plasmas générés par impulsions électriques nanosecondes répétitives dans l'air préchauffé à pression atmosphérique, Ecole Centrale Paris.
[3] Stancu G D, Janda M, Kaddouri F, Pai D, Rolon J C and Laux C O 2008, *AIAA* 2008-3882, 39th Plasmadynamics and Lasers Conference (*Seattle, WA*)
[4] Berden G, Peeters R and Meijer G 2000 *Int. Rev. Phys. Chem* **19**, 565.
[5] Laux C O, Spence T G, Kruger C H, Zare RN 2003 *Plasma Sources Sci. Technol* **12**, 125.

Journal of Physics: Conference Series **157** (2009) 012006 doi:10.1088/1742-6596/157/1/012006

In situ vehicle engine exhaust measurements of nitric oxide with a thermoelectrically cooled, cw DFB quantum cascade laser

V L Kasyutich[1], R J Holdsworth[1] and P A Martin[2,3]

[1] TDL Sensors Ltd., CTF, 46 Grafton St., Manchester, UK
[2] School of Chemical Engineering and Analytical Science, University of Manchester, Sackville St., Manchester, M60 1QD, UK
[3] Author to whom correspondence should be addressed

Email: philip.martin@manchester.ac.uk

Abstract. This paper describes the application of a thermoelectrically cooled, continuous wave (cw), single mode distributed feedback (DFB) quantum cascade laser (QCL) to the real-time, in situ vehicle engine exhaust monitoring of nitric oxide. Experiments have been carried out on a static gasoline engine test bed where the engine operating conditions can be varied in a controlled manner. Results show the response of the nitric oxide concentration to engine conditions. The prospects for further multispecies in situ measurements are discussed.

1. Introduction

Vehicle emissions are an increasing concern and give rise to a range of environmental problems related to air quality. Nitric oxide (NO) is one of the key pollutants emitted from the exhausts of both gasoline and diesel engine vehicles and is generated primarily due to high temperature combustion processes involving air as the oxidant. It is not regarded as toxic but can be rapidly oxidised to nitrogen dioxide, NO_2, which is harmful. Additionally, the NO to NO_2 conversion is involved in the production of photochemical ozone [1]. The ability to measure NO directly in the engine or in the exhaust could link into engine management systems for its control. In situ measurement could also be used to assess the condition of a catalytic converter. The traditional approach to high sensitivity measurement of NO (and NO_2) is chemiluminescence [2] but this involves extraction of the exhaust gases into the measuring unit. Infrared laser absorption spectroscopy has many advantages for this application. Previous work on vehicle exhausts has employed multiplexed near-infrared diode lasers for the detection of CO, CO_2 and CH_4 [3]. These can be made into ruggedised instruments with fibre optic cables for beam transmission. This technique has also been successfully demonstrated over many years on aero-engines by Hansen's group [4]. However, nitric oxide is only a weak absorber in the near-infrared. A room temperature mid-infrared quantum cascade laser (QCL) is a viable alternative for detection in the fundamental band. Allen's group has previously employed a pulsed (QCL) for extractive measurements in combustor exhaust flows [5]. Duxbury's group has also demonstrated the application of a pulsed QCL for extractive measurements in vehicle exhausts [6].

This paper describes the application of a mid-IR spectrometer based on a thermoelectrically-cooled cw single mode quantum cascade laser to the in situ measurement of nitric oxide directly in the exhaust of a static internal combustion engine.

2. Experimental arrangement

2.1. QCL spectrometer

The thermoelectrically cooled cw DFB QCL (Alpes Lasers) operated in the region around 1900 cm^{-1}(~5.263 μm) for the detection of nitric oxide in the fundamental band. Figure 1 shows a simulation of the nitric oxide absorption spectrum (10,000 ppm) in this region combined with the absorptions of other relevant combustion species, H_2O (5%), CO (1%) and CO_2(1%) for a temperature of 500 K and a pathlength of 23 cm. The QCL spectrometer has been described in detail previously [7] so only the essential details are given here. The main components are shown schematically in figure 2. The QCL laser light was collimated by an uncoated CaF_2. The QCL, the thermoelectric cooler, and the collimating lens were placed in to an aluminum enclosure (QCL head). The QCL current was supplied up to 0.53 A by a current driver (TDL Sensors Ltd) with a current short-term stability of less than 25 μA. The output maximum averaged power of the QCL used in the experiments was measured to be ~2.5 mW. The wavelength of the QCL was tuned by applying a modulation voltage waveform, generated by a 16-bit data-acquisition (DACQ-TDLS) card (DACQ200, TDL Sensors Ltd) with a scan rate of 952 Hz, to the current driver. The amplified signal of the thermoelectrically cooled photodetector was converted by the 16-bit analogue-to-digital converter of the DACQ-TDLS card, averaged 1000 times and then transferred, by means of a serial data connection, to a computer for further processing and reporting NO concentration. Concentrations were determined in real-time by a non-linear least-squares fitting routine. A minimum detectable optical depth of 0.0018 was estimated at SNR=1 and the NO limit of detection was 8ppm.

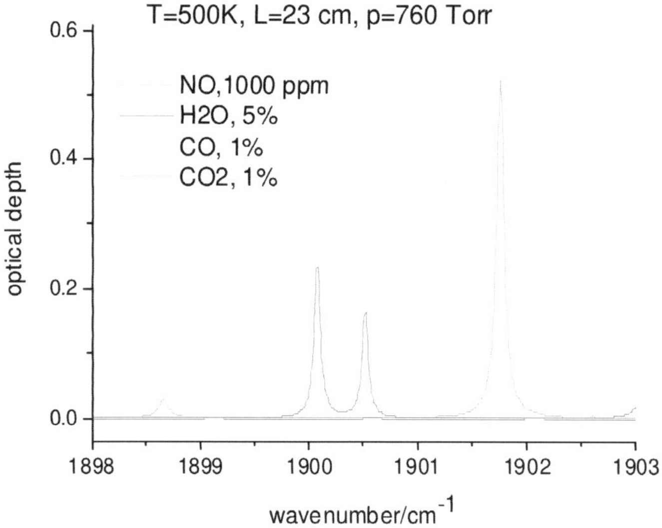

Figure 1. Simulated absorption spectra for NO, H_2O, CO and CO_2 at 500 K.

Journal of Physics: Conference Series **157** (2009) 012006 doi:10.1088/1742-6596/157/1/012006

2.2. Gasoline engine test bed

The engine used in the tests was a Rover K series 1.4 liter 16 valve version with a double overhead cam (DOHC) and a single point fuel injection system (SPi). This is shown in figure 3. The load could be varied by means of water brake dynamometer and the throttle control could be varied on an arbitrary scale of 0 to 255 with 255 corresponding to throttle wide open. The fuel used as unleaded gasoline with an octane rating of 95.

The mid-infrared QCL beam probed the 2-inch diameter stainless steel exhaust pipe 2.5 m away from the engine. The temperature and pressure of the exhaust gas were monitored just above the measurement point (see figure 2). To allow optical access to the exhaust pipe, a cross-piece was inserted with two wedged sapphire windows mounted 23 cm apart.

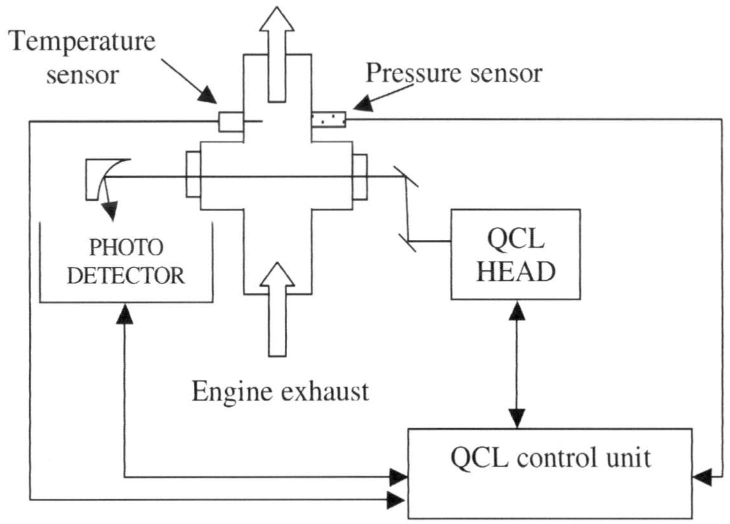

Figure 2. Experimental setup for NO measurements on engine exhaust.

Figure 3. Vehicle engine test bed. NO measurements on engine exhaust.

3. Results

Typical raw signals measured in the exhaust are shown in figure 4 for the conditions of engine on and engine off. The 20-point off period at the beginning of the scan was to allow for calibration of the 0% absorption level. The two absorption peaks in the engine on scan are the two nitric oxide absorption lines described above. The difference in intensity of the two signals was due to the attenuation of the beam due to deposits on the cell windows after the engine start.

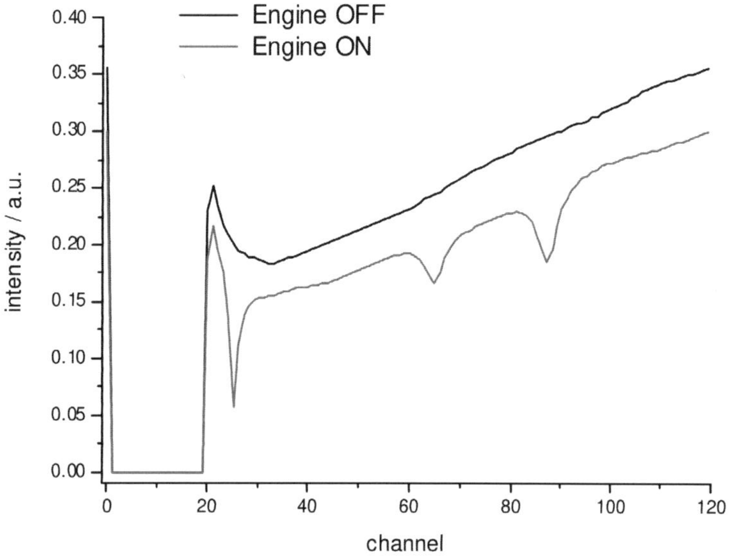

Figure 4. Typical raw signals within NO measurements on an engine exhaust (2.5 m from the engine).

Figure 5 shows the results obtained for the variation in NO concentration after engine start up. As can be seen there is a steady rise in NO emissions to over 500 ppm over a period of 15 minutes. Additionally, peak levels of up to 2000 ppm are also observed during engine warm up and engine revving. Figure 6 shows the effect on the NO concentration for a step-change in load and throttle settings. The load increased on an arbitrary scale of 40 to 75 and the throttle increased from 61 to 86 resulting in an increase of revs from 2600 to 3460 rpm and an increase in engine temperature from 550 to 650 °C. The effects of the step-change can clearly be seen with a jump in averaged NO levels from around 500 to 1750 ppm with peak concentrations up to 2000 ppm. It is also clear from the results that the NO concentrations vary considerably on a short time scale of a few seconds.

4. Discussion and Conclusions

The results presented here have demonstrated that a room temperature continuous wave (cw) DFB quantum cascade laser can be used for the in situ monitoring of vehicle exhausts. Nitric oxide concentrations have been determined in real-time and clearly respond to changes in engine operating conditions. The increases in NO levels are primarily linked to the increased combustion temperature in the cylinder due to engine startup and primarily to the increasing load on the engine. The rapid sweep integration technique employed here output NO concentrations every second (1000 averages) but this could easily be adapted to produce concentrations on the millisecond timescale. Furthermore, fixed wavelength techniques could also be used for faster measurements.

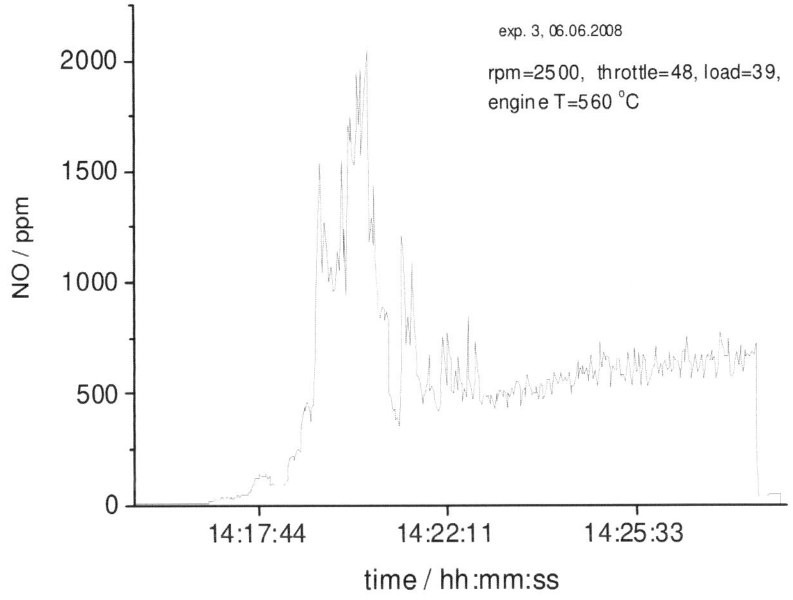

Figure 5. NO concentration variation after the engine start.

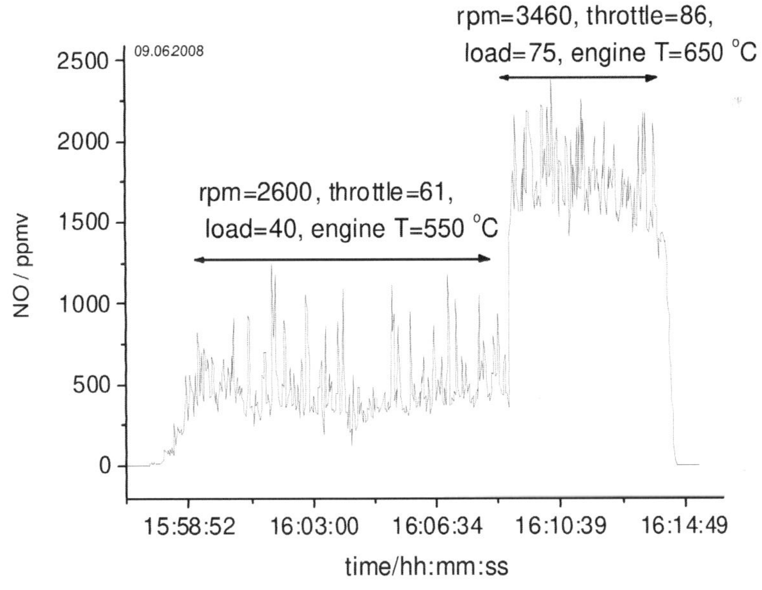

Figure 6. NO concentration variation after the engine start.

Journal of Physics: Conference Series **157** (2009) 012006 doi:10.1088/1742-6596/157/1/012006

For more detailed analysis of NO_x production under defined engine operating conditions it is planned to simultaneously monitor NO_2 concentrations in the region around 1600 cm^{-1} with another QCL operating in a multiplexed mode. It is also planned to combine the NO_x measurements with the near-infrared multiplexed spectrometer for simultaneous CO, CO_2 and CH_4 detection [3]. However, for higher sensitivity operation in the mid-infrared is preferable and CO/CO_2 detection can be accomplished with either a QCL or a difference frequency generated laser [7].

In conclusion, the prospects for highly sensitive, simultaneous multispecies, real-time monitoring directly in vehicle engines and exhausts for diagnostic and control purposes based on the latest mid-infrared laser technology appear now to be feasible.

5. Acknowledgments.

Financial support of this work was partially provided by the North West Development Agency (NWDA). We also would like to thank Mr. Michael Carroll (University of Manchester) for his help during the engine tests.

References

[1] Wayne R P 2000 *Chemistry of Atmospheres* 3rd Ed (Oxford: Oxford University Press)
[2] Clarke A G 1998 *Industrial Air pollution Monitoring* (London: Chapman and Hall)
[3] Gérard Y, Holdsworth R J and Martin P A 2007 *Appl. Opt.* **46** 3937
[4] Mihalcea R M, Baer D S and Hanson R K 1997 *Appl. Opt.* **36** 8745
[5] Jimenez J L, Koplow M D, Nelson D D, Zahniser M S and Schmidt S E 1999 *J. Air Waste Manage. Assoc.* **49** 463
[6] McCulloch M T, Langford N and Duxbury G 2005 *Appl. Opt.* **44** 2887
[7] Kasyutich VL, Holdsworth R J, Martin P A, 2008 *Appl Phys.B* **92** 271

Journal of Physics: Conference Series **157** (2009) 012007 doi:10.1088/1742-6596/157/1/012007

In situ monitoring of plasma etch processes with a quantum cascade laser arrangement in semiconductor industrial environment

N Lang[1,3], J Röpcke[1], H Zimmermann[1], A Steinbach[2] and S Wege[2]
[1]INP Greifswald, Felix-Hausdorff-Str. 2,17489 Greifswald, Germany
[2]Qimonda Dresden GmbH & Co. OHG, Königsbrücker-Str. 180, 01099 Dresden, Germany

E-Mail: lang@inp-greifswald.de

Abstract Concentrations of the etch product SiF_4 were measured online and in situ in technological etch plasmas with an especially designed quantum cascade laser arrangement for application in semiconductor industrial environment, the Q-MACS Etch. The combination of quantum cascade lasers and infra red absorption spectroscopy (QCLAS) opens up new attractive possibilities for plasma process monitoring and control. With the realization of a specific interface the Q-MACS Etch system is synchronized to the etch process and allows therefore automated measurements, which is important in a high volume production environment.

1. Introduction

Plasma technological procedures, like etching or deposition techniques play a key role for the industrial production of semiconductor devices. The challenge doing this can be found in the need of higher reproducibility when realizing structures on nanometer scales. In particular the multitude of discrete processing steps starting from the blank silicon wafer up to the final chip requires more and more methods of process monitoring to realize stable process control. Furthermore, the usage of molecular plasmas makes special demands for methods of in-situ process control.

Up to now, typically optical emission spectroscopy (OES), self excited electron resonance spectroscopy (SEERS) and mass spectrometry is applied for the online monitoring of plasma etch processes [1]. The drawback of all these methods is the lack of direct and quantitative access to etch relevant properties, like the concentration of precursors or of molecular etch products in the plasma bulk. A powerful diagnostic approach is given by infra red absorption spectroscopy with continuous sources (FTIR) [2] as well as with narrowband led salt lasers (TDLAS) [3,4] and quantum cascade lasers (QCLAS) [5-10]. The availability of quantum cascade lasers (QCL) opens up new attractive possibilities for the application of infra red absorption spectroscopy for plasma process monitoring and control. QCL emit in the MIR spectral range at room temperature, i.e. without the need of strong cooling. Because of that it is possible to design very compact MIR sources which are characterized by

[3] To whom any correspondence should be addressed.

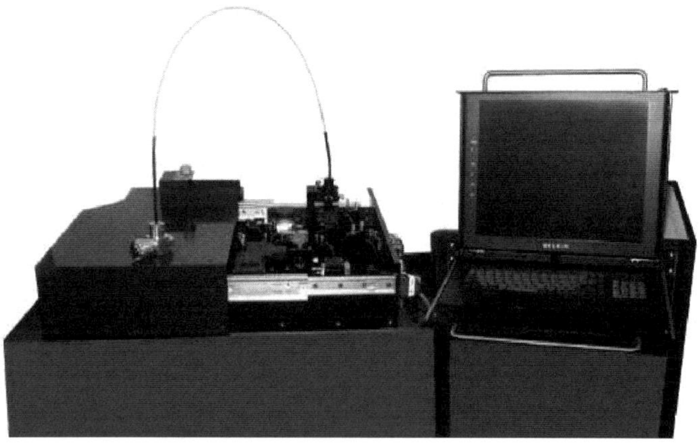

Figure 1. Picture of the Q-MACS Etch with the optical module and the separate control unit in a rack.

single mode operation and high radiance power. Sensitivities in the range of ppb with time resolutions of better than ms are realizable.

With the compact and modular measurement device Q-MACS [5] it is possible to use the potential of the QCL for plasma process control or for high sensitive trace gas analysis. It combines the advantages of absorption spectroscopy in the mid infra red spectral range with the positive properties of quantum cascade lasers. In this contribution we report actual results using the Q-MACS Etch system for the in situ process monitoring of the industrial production of DRAMs. In situ measured concentration devolutions of the etch product SiF_4 demonstrate the abilities for a process control.

2. The Q-MACS Etch

On the basis of Q-MACS a two channel spectroscopical system was developed, which fulfills the special requirements for an application in a semiconductor environment. A picture of this Q-MACS Etch can be seen in figure 1 showing the separate optical module and the control unit in a rack. Beside compactness and integrability a new concept for the optical coupling to an industrial dual frequency, capacitive coupled, magnetically enhanced, reactive ion etcher (MERIE) was developed. For simple operation and avoiding of open optical pathways an optical coupling with a MIR fiber to the etch chamber has been realized. The fiber can be mounted in front of KBr windows either for side access or

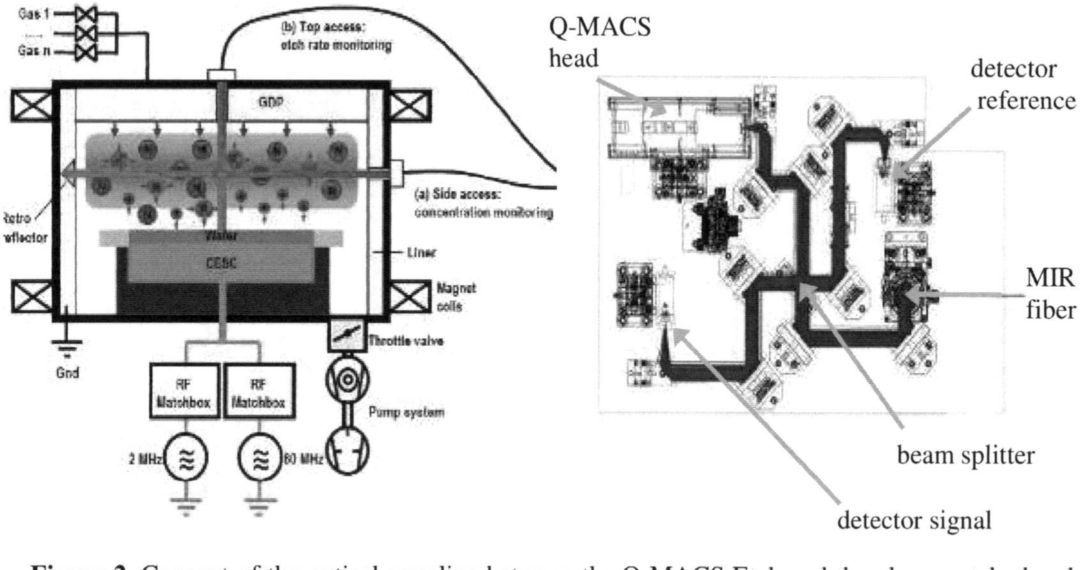

Figure 2. Concept of the optical coupling between the Q-MACS Etch and the plasma etch chamber with the schematic of the optical board.

for top access to the chamber, see figure 2 left. In case of the side access mode the infra red beam is reflected via a mirror in the liner door and coupled back into the fiber. Because of this only one optical window is necessary. This arrangement is used for the concentration measurement of etch relevant species. In case of the top access mode, the surface of the wafer itself serves as reflective element. Because of the partial transparence of silicon and silicon dioxide in the mid infra red spectral range, this arrangement can be used for an interferometric etch rate determination.

The optical setup together with the calculated beam paths via ray tracing methods is shown in figure 2 on the right. The main components consist of the laser source, the detectors, opto-mechanical parts and an MIR fiber as well. With the help of a central beam splitter both the QCL emission is separated into a measurement and a reference channel and the input MIR radiation is decoupled from the reflected signal from the chamber. In the reference channel absorption lines of the reference gas C_2H_4 are measured simultaneously in order to fix the QCL emission at a distinct spectral position with the help of a control loop (line locking). Adapted to clean room requirements and for low maintenance level thermoelectric cooled detectors are installed on board.

3. Spectroscopy at the etch product SiF_4

In the plasma process under study deep trench structures were etched in Si wafers using NF_3 as precursor. One of the main etch products is known to be the molecule SiF_4. To determine the spectral absorption characteristics for the proper selection of a QCL with its narrow spectral bandwidth, reference gas cells were filled with SiF_4 and NF_3 and were measured with a FTIR [11]. Figure 3a shows the absorption band of SiF_4 at 1030cm^{-1} due to the ν_3 stretch of the molecule. It becomes obvious that there is a spectral overlapping with the precursor NF_3. Taking this into account, the tuning of the QCL was performed so that there is a minimum of overlapping with the NF_3 absorption to minimize the errors in the concentration determination of SiF4. For the selected spectral range of 1028±0.3cm^{-1} the ratio between the absorption cross sections of both molecules is close to 90. The part of the absorption spectrum measured with the Q-MACS is shown in figure 3b. The laser line width of approximately 0.008cm^{-1} impedes to resolve spectrally the single rovibronic transitions. Using a numerical approach for the treatment of such complex spectra an effective absorption cross section of SiF_4 was determined to σ=7.7(±0.7)×10^{-18}cm^{2}molecule^{-1} and for NF_3: σ=8.7(±0.8)×10^{-20}cm^{2}molecule^{-1} [11]. With the experimentally determined absorbance and the known absorption length the concentration of the absorbing species follows from the Lambert-Beer law. Because of the broad band absorption characteristic of SiF_4 in the tuning range of the QCL it was necessary to measure the

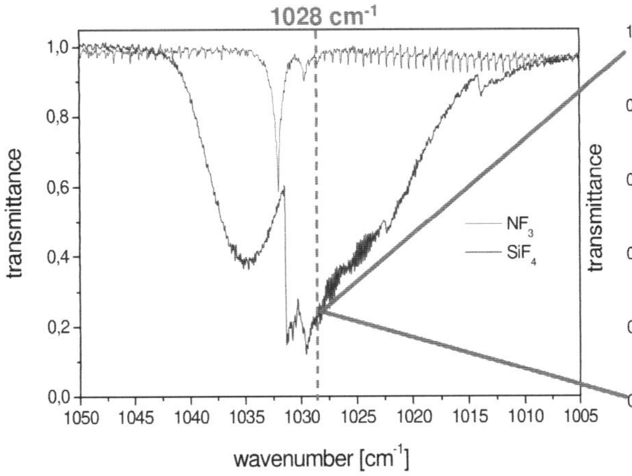

Figure 3a. FTIR spectra of SiF_4 and NF_3 from 1005cm^{-1} to 1050cm^{-1}, the spectral resolution is 0.11cm^{-1} [11].

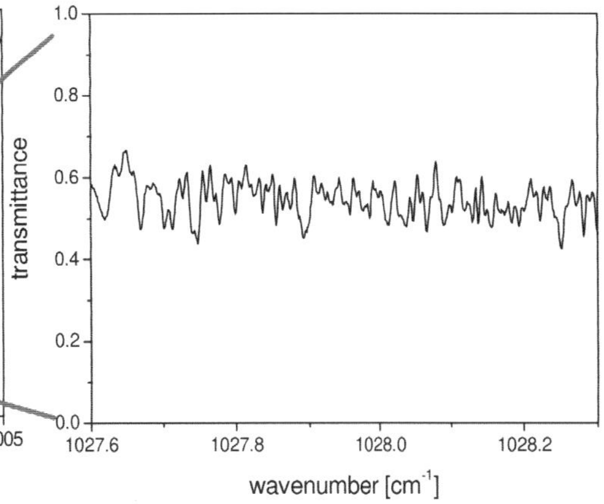

Figure 3b. QCLAS spectrum of SiF_4 from 1027.6cm^{-1} to 1028.3cm^{-1}, instrumental broadening is 0.008cm^{-1} [11].

background signal for the reference signal separately.

4. Synchronization with the etch tool

The above discussed absorption behavior of SiF_4 impedes the simultaneous detection of the background intensity of the QCL during the measurement. Having a look on the time course of the plasma etch process different states in the chamber are distinguishable and can be related to different phases of the process. This can be used for the automated determination of the concentration, as illustrated exemplarily in figure 4 for a conditioning sequence. If the liner door is open the QCL radiation is no longer reflected back from the chamber and the detected signal corresponds to an offset I_{off}, which is important to take into account in the calculation. As soon as the liner door is closed the background intensity I_0 is measured before the process gases starts to flow in the stabilization step. During the plasma phase (ME) the signal $I(t)$ corresponds to the actual transmission. At the end of the plasma process again it is possible to measure the background intensity I^0_2, which can be altered because of depositions at the window or mirror. Therefore the concentration N at a time t is given by:

$$N \propto \int d\nu \ln \frac{I_i^0 - I_{off}}{I(t) - I_{off}}$$

To be able to use these different meanings of the signal a synchronization between the Q-MACS Etch and the etch chamber control tool is required. This was realized by means of a communication via a customized network interface. With the exchange of attributes it was possible to interpret at any time point of the process the signal and therefore to calculate the concentration. From the changes of the background intensities I^0_i a degradation coefficient was determined in a linear approach and the concentration corrected correspondingly. The effect of the degradation on the concentration can be seen in figure 5 exemplarily. On the left side the signal course is shown as recorded from the measurement detector from batch processed wafers with the corresponding degradation coefficient. The sequence was measured during the etching of 25 masked wafers for deep trench structures after seasoning the chamber. Every etching process was followed by a cleaning cycle consisting of a special clean process for the cathode (ICClean) and a cleaning of the chamber (DryClean). In the right graph of figure 5 both SiF_4 concentration values are shown, the uncorrected in grey and with degradation taken into account in black. During the etching processes the differences become significant whereas the effect is less pronounced during the cleans.

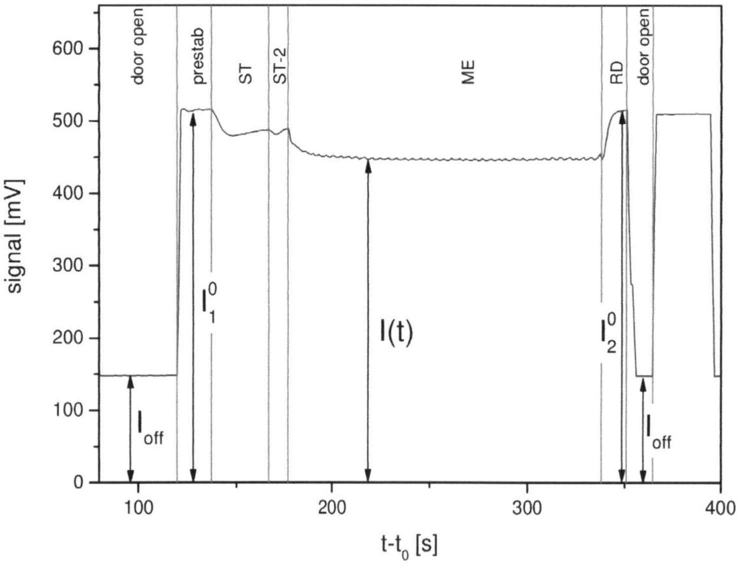

Figure 4. Time course of the measurement signal $I(t)$ during a seasoning sequence where I_{off} is the offset, I^0_1 and I^0_2 the background before and after the process. The different meanings according to the actual status of the process are given: prestab, ST and ST-2 – stabilisation steps, ME – main etch, RD – ramp down and door open.

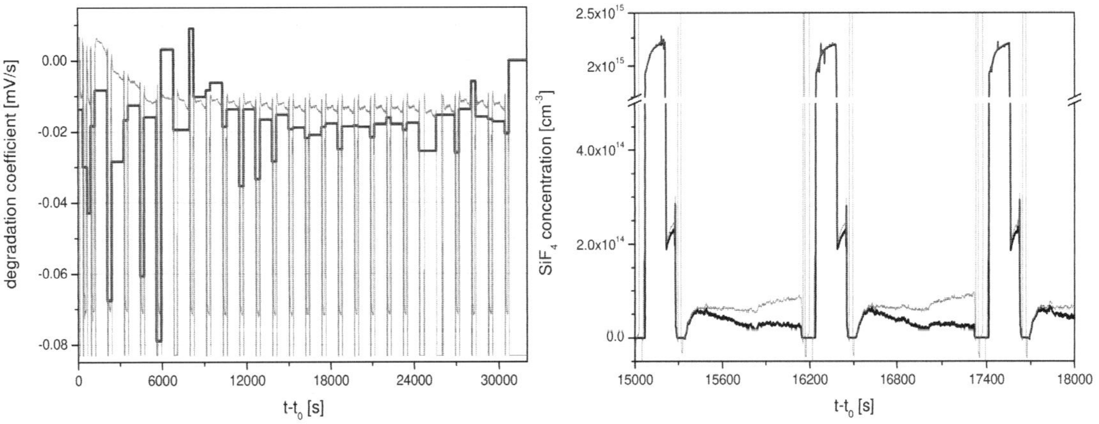

Figure 5. Left: course of the degradation coefficient (blue) and the measurement signal (grey) during the processing of 25 wafers. Right: the SiF_4 concentration values with degradation taken into account (black) and without (grey).

With this the concentrations of the etch product SiF_4 can be transferred to the tool and used for process monitoring or control purposes. The synchronization with the etch tool fulfils therefore an important requirement for the online process monitoring with Q-MACS in typically automated high volume production environments.

5. Monitoring of the Si etch process

In the experiment reported here 25 masked Si wafers were processed for the plasma etching of deep trench structures in a batch mode. Between the etching sequences the chamber was cleaned with an ICClean and a DryClean procedure. Before the etching procedure the chamber was conditioned twice via a process-like plasma and blank Si wafers. Taken into account the above discussed degradation the resulting concentration courses were like in figure 6 presented. The individual traces were overlaid and normalized with respect to their starting time in order to compare the devolution of the SiF_4 concentration for each wafer. For reason of clarity all 25 graphs are split into two graphs. As already recognizable in the signal traces in figure 5 (left) the development of the SiF_4 concentrations during the deep trench etch processes are affected by a seasoning which is not present in the subsequent clean cycles, see figure 7. Furthermore, for the latter the SiF_4 concentrations were detected stable and

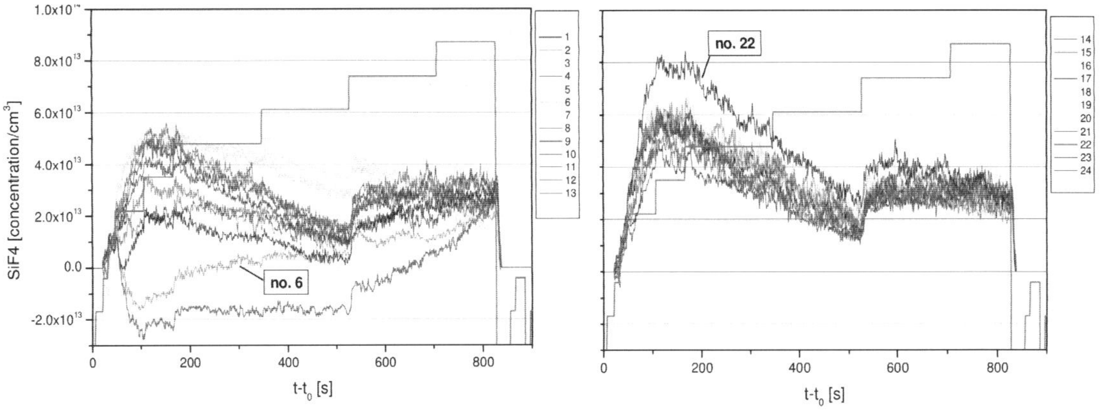

Figure 6. SiF_4 concentration devolutions during the individual etch processes. The first 13 process courses are shown on the left, the remaining on the right with same scaling. In addition the single process steps are emphasized in grey.

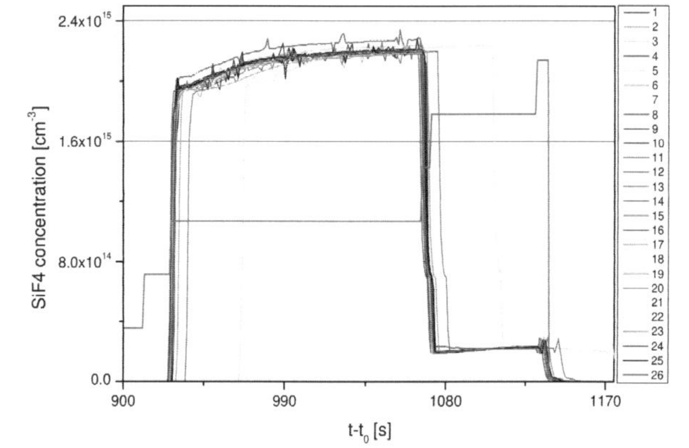

Figure 7. Development of the SiF_4 concentration during the DryClean cycles.

reproducible since the beginning of the batch. There is a time jitter in the concentration values due to different spans between two successive processes caused by the tool control. In case of the deep trench processes it takes up to 8 cycles before the concentration courses becomes more reproducible. This is seen as a clear indication of chamber conditioning effects, i.e. the balancing of chemical reactions is dominated by plasma wall interactions. In dependence of the prehistory it takes several process cycles until stable conditions prevail in the etch chamber. Because of that the measured concentration courses show a stronger scattering at the beginning of the experiment. Moreover, also negative values were detected for the SiF_4 concentration, which is an artifact and does not have any physical meaning. The appearance is related to an enhanced background during the process. With increasing number of processed wafers the concentration courses become more and more reproducible as it can be seen in the right graph in figure 6. During the etch process SiF_4 is built up to $6\times10^{13}cm^{-3}$ which corresponds to a partial pressure of ca. 1%. Significant different concentration values were detected for the wafer no. 22 with about 30% higher SiF_4 concentrations in the first fourth of the etch process. As in case of wafer no. 6 the previous cleaning procedure failed and was stopped from the tool control. Again the amount of detectable SiF_4 during the etch process depends sensitive on the chamber conditioning.

In figure 8 first results of inline data of the 25 processed wafers are presented. The etch results are evaluated via an optical method in the infra red spectral range and statistically analyzed with the help of a complex model. The deep trench structures are characterized besides other properties by the trench depth and the bottom void. The latter is a measure for the area at the bottom of the trench and for the roughness at the trench side wall. The results for the trench depth show a homogenous distribution around 6.9µm except small deviations in case of wafer 6, 21 and 22, see left graph in figure 8. For these wafers the trench depths are significant up to 200nm smaller. The values for the

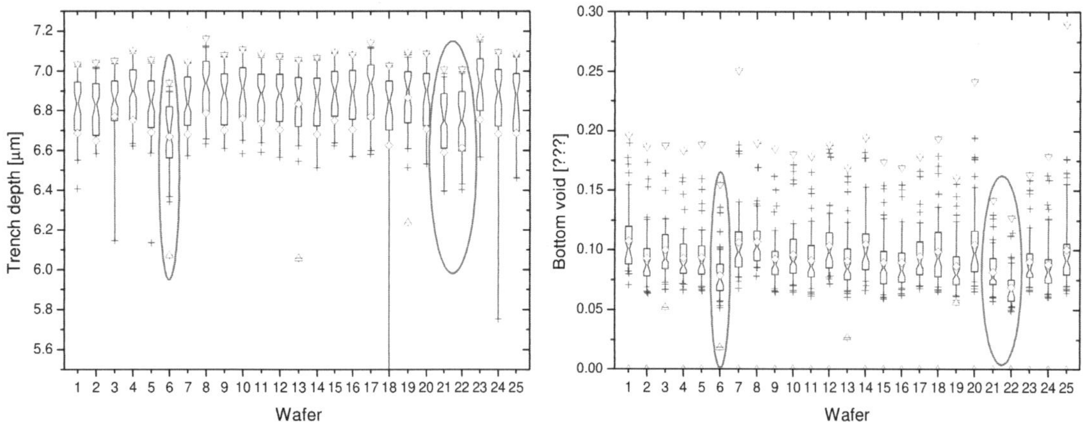

Figure 8. Boxplot of the first inline data: deep trench etch depth (left) and bottom void (right).

bottom void show an analog distribution as the trench depth, as it can be seen on the right graph. Again the influence of the failed clean cycles becomes apparent.

6. Summary

A quantum cascade laser system Q-MACS Etch was developed for process monitoring in the semiconductor industry environment. For the first time one of the major etch products SiF_4 was monitored in situ in silicon plasma etch processes used for the production of DRAM modules. Based on the determination of an effective absorption cross section it was possible to measure online the molecular SiF_4 concentration during the process. In order to evaluate the reproducibility and the stability of the plasma processes the measurements were expanded to a batch processing of 25 silicon wafers. The required synchronization with the etch process was realized via a customized network interface. With this the Q-MACS Etch communicates with the etch tool providing the automation of data acquisition. The usage of a special MIR fiber in combination with an internal mirror is an innovative approach to ensure robustness and flexibility in an industrial production environment.

The first application of a quantum cascade laser arrangement for monitoring of industrial etch processes has opened up a challenging new option for control of demanding semiconductor production. Focused on sensitive and fast concentration measurements of molecular key components while ensuring compactness, robustness and long term stability, this new class of process control equipment has the potential to become implemented also into other fields of plasma technology.

Acknowledgements

The work for his paper was supported within the scope of technology development by the EFRE fund of the European Community and by funding of the State Saxony of the Federal Republic of Germany.

K.-D. Weltmann is gratefully acknowledged for his continuous encouragement and general support making the present work possible. The authors are especially thankful to S. Barth, M. Reinicke, S. Herberg and U. Macherius for their permanent support and stimulating climate. We are also indebted to F. Hempel, S. Glitsch and F. Weichbrodt for the technical assistance and useful discussions.

References

[1] Steinbach A, Barth S, Henke A, Wege S, Sobe J, Reinicke M and Kijowski M 2005 *6-th European Advanced Equipment Control / Advanced Process Control (Dublin, Ireland)*
[2] Grählert W, Dani I, Throl O, Hopfe V, Pietsch K, Wünsche T and Dreyer T 2006 *VDI-Berichte* **Nr 1959** 45
[3] Heason D J, Spencer A G 2003 *J. Phys. D: Appl. Phys.* **36** 1543
[4] Sun H C, Patel V, Singh B, Ng C K and Whittaker E A 1994 *Appl. Phys. Lett.* **64**(21) 2779
[5] Hempel F, Glitsch S, Röpcke J, Saß S and Zimmermann H 2005 *Plasma Polymers and Related Materials* ed M Mutlu 142
[6] Röpcke J, Glitsch S, Hempel F, Saß S, Schulz K-D, Weltmann K-D and Zimmermann H 2005 *Proc. 48th Annual SVC Conf.* (Denver, USA) 54
[7] Röpcke J, Lombardi G, Rousseau A and Davies P B 2006 *Plasma Sources Sci. Technol.* **15** 148
[8] Röpcke J, Glitsch S, Hempel F, Lang N, Macherius U, Saß S, Stancu G D, Weichbrodt F, Weltmann K-D and Welzel S 2006 *VDI-Berichte* **Nr 1959** 279
[9] Welzel S, Gatilova L, Röpcke J and Rousseau A 2007 *Plasma Sources Sci. Technol.* **16** 822
[10] Röpcke J, Welzel S, Lang N, Hempel F, Gatilova L, Guaitella O, Rousseau A and Davies P B 2008 *Appl. Phys. B* **92** 335
[11] Stancu G D, Lang N, Röpcke J, Reinicke M, Steinbach A and Wege S 2007 *Chem. Vap. Deposition* **13** 351

Time resolved QCLAS measurements in pulsed cc-rf CF_4/H_2 plasmas

S Stepanov[1], S Welzel[2], J Röpcke[2] and J Meichsner[1]

[1] Institute of Physics, University of Greifswald, Felix-Hausdorff-Str. 6, D-17489, Greifswald, Germany

[2] Leibniz Institute for Plasma Science and Technology, Felix-Hausdorff-Str. 2, D-17489, Greifswald, Germany

E-mail: stepanov@physik.uni-greifswald.de

Abstract. Fluorocarbon containing capacitively coupled radio frequency (cc-rf) plasmas are widely used in technical applications and as model systems for fundamental investigations of complex plasmas. Absorption spectroscopy based on pulsed quantum cascade lasers (QCL) was applied in the mid-IR spectral range of 1269-1275 cm^{-1}. Absolute densities of the precursor molecule CF_4 and of the stable product C_3F_8 were measured with a time resolution of up to 1 ms in pulsed CF_4/H_2 asymmetrical cc-rf (13.56 MHz) discharges. For this purpose both the non-negligible temperature dependence of the absorption coefficients and the interference of the absorption features of CF_4 and C_3F_8 had to be taken into account in the target spectral range. Therefore, at two different spectral positions composite absorption spectra were acquired under the same plasma conditions in order to discriminate between CF_4 and C_3F_8 contributions. A total consumption of ~ 12 % was observed for CF_4 during a 1 s plasma pulse, whereas C_3F_8 appeared to be produced mainly from amorphous fluorocarbon layers deposited at the reactor walls. A gas temperature increase by ~ 100 K in the plasma pulse was estimated from the measurements. Additionally, not yet identified unresolved absorption (potentially from the excited CF_4 molecule) was found during the 'on-phase'.

1. Introduction

Manifold applications of fluorocarbon plasmas and their increasing importance in industry motivate fundamental investigations of these complex plasmas. Pulsing of the discharge can be used as an additional parameter for process control, since many plasma parameters, e.g. densities and temperatures, become time dependent [1, 2]. Therefore, absolute concentrations of key stable and transient species measured in pulsed plasmas with a high time resolution are essential for a better understanding of the plasma chemistry and kinetics in these plasmas.

Within the last decades absorption spectroscopy (AS) has been established as a suitable diagnostic tool for monitoring absolute ground state number densities in plasmas. For years tunable lead salt diode lasers (TDL) have been used as light sources in the mid infrared (IR) spectral range between 3 and 20 µm, where characteristic absorption lines of many molecular species are normally located [3, 4]. In particular infrared tunable diode laser absorption spectroscopy (IR-TDLAS) has been successfully applied for time resolved studies of CF-, CF_2-radicals and C_2F_4 as a stable reaction product in capacitively coupled radio frequency (rf) CF_4/H_2 pulsed plasmas (13.56 MHz) [1].

Furthermore, based on the simultaneous monitoring of selected absorption lines, a novel method was developed for the time resolved determination of the rotational temperature of the CF_2 radical within a single plasma pulse [2]. Recently, quantum cascade lasers (QCL) have become available as alternative light sources in the mid-IR spectral range [5, 6] Distributed feedback (DFB)-QCLs combine narrow bandwidth single-frequency operation with continuous tunability over several wavenumbers, and average powers over a mW. Hence they are not only used for highly sensitive trace gas spectrometers in various fields of applications, e.g., isotope measurements [7], atmospheric sensing [8], explosives detection [9] or breath analysis, but also being increasingly employed for plasma diagnostics [10-12] or plasma process monitoring [13]. In particular pulsed QCLs offer the opportunity to achieve a much higher time resolution compared to lead salt laser systems [11]. If a relatively long QCL pulse of a few 100 ns is applied a full absorption spectrum of up to a few cm^{-1} can be acquired within the laser pulse [14]. The time resolution is thus theoretically limited to the pulse length of the laser.

In this contribution a study on pulsed CF_4/H_2 capacitively coupled rf discharges is reported focussing on the time resolved measurements of the absolute concentrations of stable molecules. Therefore quantum cascade laser absorption spectroscopy (QCLAS) employing a pulsed room-temperature QCL emitting at 7.86 μm has been applied. The QCL spectrometer was co-aligned to an existing TDL setup [2] and time resolved concentration measurements have been performed on CF_4 and C_3F_8. The experimental arrangement and the data handling approach are described in section 2. The results are presented and discussed (also in terms of the alternating gas temperature in the discharge) in section 3 and summarised in section 4.

2. Experimental set-up and data handling

2.1. Discharge set-up

The experimental arrangement is shown schematically in figure 1. The discharge was created in a cylindrical stainless steel vacuum chamber with a total volume of ca. 20 l. A cylindrical copper block of about 8 cm in diameter was placed in the centre of the chamber and capacitively coupled with the 13.56 MHz rf power generator (ENI ACG-5) by means of a matching network (ENI MW-5). The chamber walls were grounded resulting in a highly asymmetrical rf discharge. The power generator was pulsed by a master pulse generator (BNG 555) with a period of 3 s. The process gases CF_4 and H_2 were fed into the reactor at a flow rate of 7 and 3 sccm, respectively, through small holes in the shielding around the powered electrode. The total pressure in the chamber (5-25 Pa) was measured with a capacitance manometer (MKS pressure transducer type 122A) and controlled independently from the constant gas flow rates. For that the pumping speed of the rotary pump (Leybold, D 25 BCS) was regulated by means of a throttle valve controller (MKS pressure controller type 652) which also enabled the pressure to be recorded with a time resolution of ~ 0.25 s.

2.2. Spectroscopic set-up

The QCLAS measurements of CF_4 and C_3F_8 were carried out in the spectral range of 1269-1275 cm^{-1} using a pulsed thermoelectrically (TE) cooled QCL (Alpes Lasers) housed in a QCL measurement and control system (Q-MACS, neoplas control) [15]. The QCL was operated with 300 ns pulses at two different heat sink temperatures and therefore provided spectral scans centered around 1271 cm^{-1} and 1274 cm^{-1}, each of ~ 1 cm^{-1} spectral coverage [16]. The light emitted by the QCL was guided in a double pass through the reactor approximately 4 cm above the rf electrode (figure 1). The total absorption path length through the plasma chamber was ~ 90 cm. The transmitted signal was focused onto a fast TE cooled detector (PDI-3TE-10/12, VIGO) followed by a fast preamplifier (neoplas control) and a fast digitizing oscilloscope (WR 104Xi, LeCroy). In order to generate a stable trigger for the data acquisition and to avoid any jitter between the driving electronics and the optical output, a small portion of the QCL beam was deflected using a ZnSe beam splitter onto a second fast TE cooled detector in a reference path. The rising edge of the reference signal served as trigger for the acquisition of the QCL signal transmitted through the reactor.

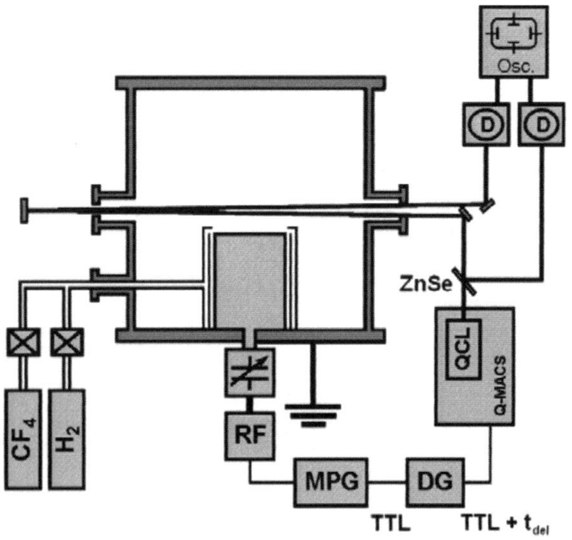

Figure 1. Experimental set-up (MPG - master pulse generator, DG - delay generator, RF – rf power generator, D – detector with preamplifier, Osc. - oscilloscope).

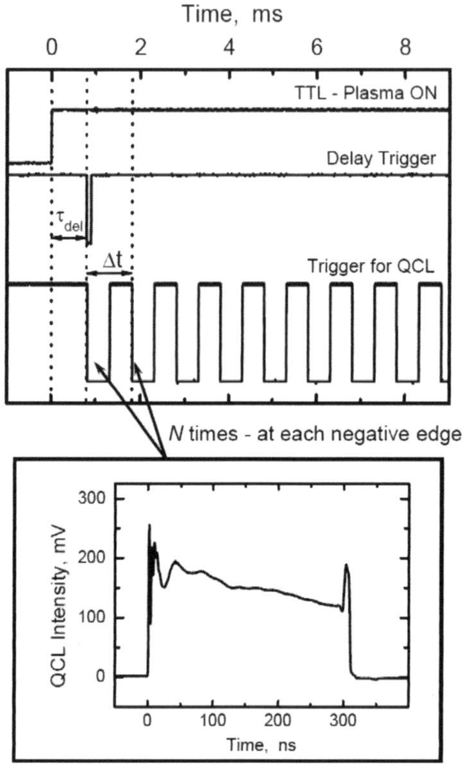

Figure 2. Trigger scheme for the time resolved data acquisition after igniting the discharge. *Upper:* A master (TTL) pulse triggers the discharge and *N* subsequent QCL pulses (via a frequency generator). *Lower:* Sample spectrum of a single QCL pulse.

To synchronize the QCL pulses with the discharge, a gate TTL trigger of the master pulse generator was used (figure 2). This master gate signal determined the 'plasma-on' phase and realized simultaneously the master trigger for all subsequently recorded QCL spectra relative to the plasma pulse. A delay generator launched a sequence of N short trigger pulses for the QCL with a delay time of τ_{del} after the master gate signal (figure 2). This burst of the QCL trigger events was realized by a frequency generator (Hewlett Packard). The repetition frequency of these N short pulses defined the time resolution of the measurements Δt. Thus, the sampling of the plasma pulse by means of QCL pulses was determined by three parameters τ_{del}, N and Δt, respectively. Typically, a burst of $N = 600$ pulses with a time resolution of $\Delta t = 5$ ms, launched $\tau_{del} = 500$ ns after the master gate signal, was used to map the entire plasma pulse. For a more detailed analysis of the plasma ignition or early afterglow a better time resolution of $\Delta t = 1$ ms was chosen. Finally, to increase the signal to noise ratio, 25 of such sequences (i.e. from 25 different plasma cycles) were recorded and averaged in a first step of the post data treatment yielding N averaged spectra.

For further data analysis the signal without any absorbing species (base line) and with a germanium etalon (25.4 mm, 0.0485 cm^{-1} free spectral range) in the beam path were recorded after each series of measurements. N_2O and CH_4 were used as reference gases for absolute calibration of the frequency scale.

2.3. Data handling

The number density n of the absorbing species can be derived from the relationship between the incident (I_0) and transmitted (I) laser intensity, based on Beer-Lambert's law:

$$I(\nu) = I_0(\nu)e^{-n\sigma(\nu,T)L} \tag{1}$$

where L is the absorption length and $\sigma(\nu,T)$ is the temperature dependent absorption cross section at wavenumber ν. The complex spectra of CF_4 and C_3F_8 measured within the target spectral range at the current low pressure conditions were found to be affected by the rapidly chirped QCL pulses (so called 'rapid passage effect', [17]). Therefore calibration of the absorption cross sections was required. Following an approach similar to Stancu *et al* [13] an effective integrated absorption cross section σ^{eff} was defined:

$$\sigma_{eff} = \frac{1}{n \cdot L}\ln\frac{\langle I_0(\nu)\rangle}{\langle I(\nu)\rangle} = \frac{1}{n \cdot L}\ln\left(\frac{A_0}{A}\right) \tag{2}$$

where $\langle ...\rangle$ (further denoted as A) is the signal averaged over a distinct narrow spectral range. Further details on the σ_{eff} definition and calibration procedure including a detailed analysis of the temperature dependence $\sigma_{eff}(T)$ can be found in [16].

Since the spectral features of CF_4 and C_3F_8 (further denoted with index $k = 1$ and 2, respectively) were found to interfere in the target spectral range, the number density n_k of one of these molecules could not directly be determined from a single plasma measurement. Each acquired spectrum might be composed of contributions from both molecules. However, the significantly different dependence of the CF_4 and C_3F_8 absorption cross sections on ν provides a means to discriminate between the two species of interest. Therefore, time resolved measurements were carried out under identical plasma conditions at two separate spectral positions ν_j around 1271 cm^{-1} and 1274 cm^{-1}, respectively, where the ratio of the absorption cross sections σ_k^{eff} for CF_4 and C_3F_8 was considerably different [16]. After the initial averaging over 25 plasma cycles the time dependent absorption $I(\nu_j,t_i)$ at the considered spectral positions is obtained. Next, for each time step t_i a system of two nonlinear equations had to be solved numerically using the $\sigma_k^{eff}(\nu_j,T(t_i))$ values calculated for the current temperature $T(t_i)$ in the plasma (cf. Appendix). Finally, this yielded the time dependent absolute number densities $n_k(t_i)$ for CF_4 and C_3F_8 under considered plasma conditions:

$$I(\nu_j,t_i) \xrightarrow{\sigma_k^{eff}(\nu_j,T(t_i))} n_k(t_i) \quad (t_i = 1...N). \tag{3}$$

3. Results and discussion

3.1. Properties of the pulsed discharge

For further interpretation of the performed absorption measurements external discharge parameters were investigated in preliminary experiments. Figure 3 shows a typical rf voltage sample at the beginning of the 'plasma-on' phase obtained by averaging over 25 successive plasma cycles. Typically, ~ 1 ms after the initial trigger to the rf power generator a short and unstable plasma burst was observed, followed by ~ 0.5 ms where the plasma was absent. A second onset, ~ 1.5 ms after the initial trigger, led to stable discharge conditions. Thus, experiments with the best time resolution were carried out with $\Delta t = 1$ ms. For all other cases $\Delta t = 5$ ms was chosen.

The typical behaviour of the total pressure during a plasma cycle is shown in figure 4. The response time of the throttle valve controller was not fast enough to keep the pressure inside the chamber at a constant level. Furthermore the time resolution of the data acquisition from the controller was limited to 0.2 s resulting in a flattened slope of the pressure at the transition points. Nevertheless, the pressure increase of about 30 % can qualitatively be understood as an upper limit of the temperature increase during the 'plasma-on' phase compared to the room temperature level in the 'plasma-off' phase. This approximation leads to a gas temperature of ~ 390 K in the plasma which was used for all further calculations. Moreover, this estimation correlates with rotational temperature values found under similar conditions for CF_2 radicals in plasmas [2].

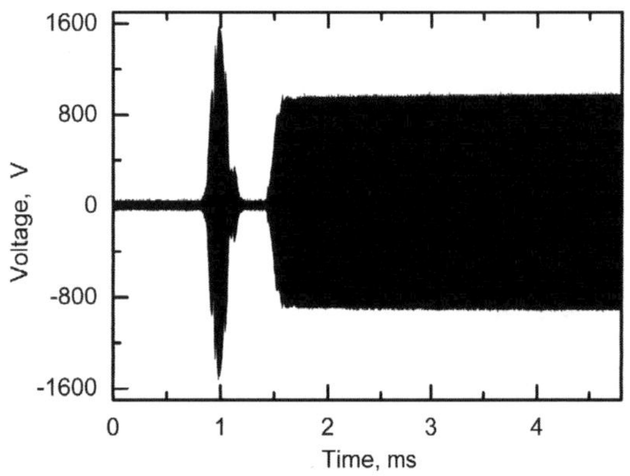

Figure 3. Typical rf sample measured at the beginning of the 'plasma-on' phase (10 Pa, 7 sccm CF_4 + 3 sccm H_2, 100 W).

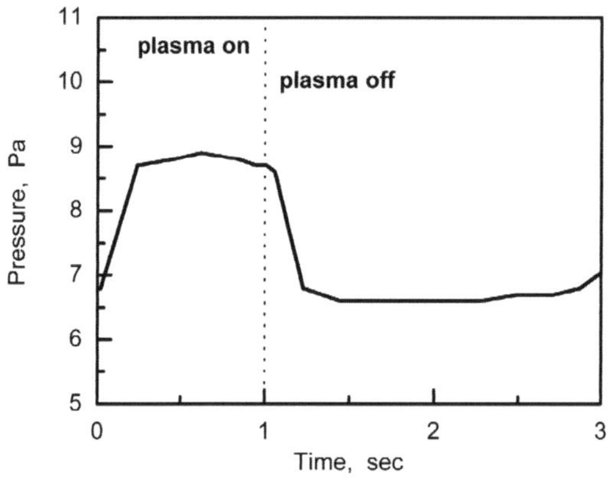

Figure 4. Total pressure change measured over a full plasma cycle (7 sccm CF_4 + 3 sccm H_2, 100 W).

3.2. Time resolved analysis of species concentration

Figure 5 shows an example for the absorbance $\ln(A_0/A)$ measured at the two studied spectral positions ν_j under identical plasma conditions. In both time traces a steep increase of the absorption in the 'plasma-on' is observed, whereas the behaviour during the plasma pulse is slightly different: at 1274 cm^{-1} where CF_4 has its strongest absorption cross section for the two positions ν_j the absorption is gradually decreasing during the plasma pulse (curve b) whereas at 1271 cm^{-1} the absorption is almost constant (curve a). If only CF_4 features were considered a decrease of the absorption values due to the conversion of the precursor gas would be expected. In the 'plasma-off' phase the absorption increases slightly at 1274 cm^{-1}, but is almost constant at 1271 cm^{-1}. Thus this example demonstrates that the individual temperature dependences of the absorption cross sections and contributions from plasma produced species to the absorption have to be taken into account for the further analysis. More detailed measurements with a higher time resolution (not shown here) at the transition points between 'plasma-on' and 'off' phase also revealed that: i) the ~ 1 ms delay of the power input into the plasma after the trigger (figure 3) was also present in the IR data, and ii) the rapid changes in the absorption at the beginning and the end of the plasma pulse occurred mostly within 25 ms.

Potential transient and stable molecules that might be detected in the CF_4/H_2 plasma are CF, CF_2 CF_3 as well as CHF_3, C_2F_4, C_2F_6 and C_3F_8. Higher fluorocarbon molecules were not very likely to be produced and present in the gas phase. Except for C_3F_8 none of the mentioned stable $C_xH_yF_z$ molecules showed absorption features in the target spectral range. CF_2 exhibits no absorption lines in the considered range. The absorption features of CF and CF_3 would be too weak to cause such a steeply rising edge as in figure 5. Moreover, the absorption lines of the CF_x radicals are typically well-resolved in a spectrum. As a consequence, $\ln(A_0/A)$ should have different values for distinct spectral micro-windows ν_j depending on whether a CF_x radical line is included or not. However such behaviour was not observed. Hence, the measured broadband absorption in the target spectral range is expected to be caused by CF_4 and C_3F_8.

Following equation (3) and considering the temperature dependence of the absorption cross sections the data handling procedure yielded the number densities of CF_4 and C_3F_8. The influence of the gas temperature in the 'on-phase' on the calculated number densities was checked separately to study their sensitivity on this input parameter. For a temperature range of 350 to 430 K no significant difference in the general trends was found. However, the value of 390 K which was estimated in section 3.1 fits best because any sharp change in n_1 is absent in this case at the 'on-off' transition. The number densities of CF_4 and C_3F_8 obtained for $T = 390$ K are shown in figure 6. A gradual increase of CF_4 in the order of 1.2×10^{14} cm^{-3} is observed during the 'plasma-off' phase. This can be attributed to the unfinished gas replacement because the residence time was estimated to be ~ 12 s for the present experimental conditions. In the 'on-phase' the initial precursor concentration decreases by ~ 12 % due to both dissociation and conversion processes. Since an additional underlying absorption of C_3F_8 was assumed for the data analysis, an immediate increase of its density n_2 from zero (or noise) level in the 'off-phase' up to 2.8×10^{14} cm^{-3} during the discharge is deduced.

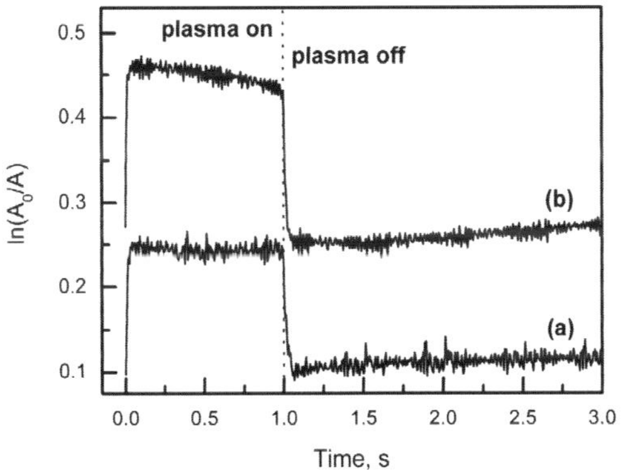

Figure 5. The ratio of the signal intensities $\ln(A_0/A)$ measured in pulsed plasma at: (a) 1271 cm^{-1} and (b) 1274 cm^{-1} (10 Pa, 7 sccm CF_4 + 3 sccm H_2, 100 W).

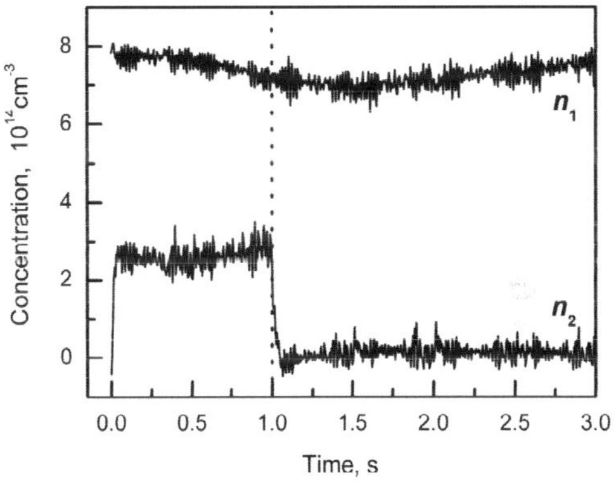

Figure 6. Time resolved absolute number densities n_1 (CF_4) and n_2 (C_3F_8) obtained for $T = 390$ K (10 Pa, 7 sccm CF_4 + 3 sccm H_2, 100 W).

The carbon-fluorine content which follows from the calculated number densities of the precursor molecule CF_4 and the product C_3F_8 (figure 6) clearly exceeds the initial content in the feed gas flow. This suggests an additional non-negligible fluorocarbon source and its conversion in the gas phase. Thus, amorphous fluorocarbon (a:C-F) layers deposited at the chamber walls during previous experiments may mainly contribute to the C_3F_8 formation in figure 6. After switching the plasma off the production via this channel stops. Diffusion losses and subsequent chemical reactions lead therefore to the sudden drop of n_2. However, the time scale of less than 100 ms observed for this decrease in n_2 is almost one order of magnitude smaller than typical values found for other stable products [1]. Moreover, if the calculated number density of C_3F_8 were present, the gas phase would also contain a significant number of C_2F_6 molecules [18, 19] which in turn would further increase the excess of fluorocarbons compared to the initial feed gas flow. Further effects or species may therefore contribute to the observed time trace of n_2.

In order to identify the main stable products and estimate their number densities, FTIR spectra of the 'off-phase' gas composition were recorded and compared with reference spectra which were previously acquired for calibration purposes. A glass cell attached to the plasma reactor was used to extract a sample of the gas phase after discharge operation. Next, the measurements were performed *ex-situ* by means of a FTIR spectrometer (Bruker Vertex 80V). Apart from the precursor molecule CF_4 the stable compounds C_2F_6, C_3F_8, CHF_3 and HF were identified. Although these complimentary measurements cannot directly be compared with the QCLAS results, they provide a means to assess the trends found in the time-resolved QCL experiments. The number densities of C_2F_6 and C_3F_8 in the sample cell were determined from a linear fit to calibration spectra and found to be of 1.8×10^{14} cm^{-3}

Journal of Physics: Conference Series **157** (2009) 012008 doi:10.1088/1742-6596/157/1/012008

and 0.4×10^{14} cm^{-3}, respectively. Since the present C_3F_8 detection limit estimated from the peak-to-peak noise level is of the same order (0.3×10^{14} cm^{-3}), the C_3F_8 density during the 'off-phase' is just below this value which also explains the observed "zero-signal" of n_2. Furthermore, if the C_2F_6 estimate is considered in respect to the calculated 'on-phase' value for n_2 (which is almost twice as big) and its time dependence, another short-lived molecule or absorption feature should be included in the data analysis. Since potential radicals exhibit no (broad-band) absorption in the present spectral range, CF_4 hot bands may be the most reasonable absorption feature [18]. Unfortunately, such an influence cannot be calibrated and included in the analysis using the present methods. Further investigations are therefore necessary to discriminate the obviously non-negligible contribution of excited CF_4 from other stable molecules of interest.

4. Summary and conclusions

The absolute number densities of CF_4 and C_3F_8 have been measured in pulsed CF_4/H_2 rf plasmas by means of time resolved QCLAS with a time resolution of typically 5 ms. In order to deconvolve the composed absorption spectra with main contributions from at least two stable species with broadband, unresolved absorption, two spectra had to be acquired under the same plasma conditions in two separate spectral micro-windows. A sensitivity analysis of the data retrieval procedure confirmed the gas temperature value of 390 K which was approximated initially from the pressure increase in the 'plasma-on' phase. The concentration of the CF_4 precursor molecule decreased by ~ 12 % during the 'on' phase, whereas the suggested C_3F_8 concentration clearly increased above the noise level and immediately followed the discharge pulse. In particular the latter fact is unexpected, since, even under the assumption that C_3F_8 might be formed in plasma from amorphous fluorocarbon layers at the chamber walls, the observed rapid decrease of C_3F_8 in the 'off-phase' was not expected for a stable molecule.

Accompanying FTIR test measurements revealed the presence of C_3F_8 in the 'off-phase' gas composition, but in the order of the QCLAS detection limit (~ 0.3×10^{14} cm^{-3}). By means of applied methods it is not feasible to distinguish between C_3F_8 and other not yet considered molecules (e.g. CF_4 hot bands) during the 'on-phase'. However, the C_2F_6 number densities resulting from the FTIR spectra analysis and the carbon-fluorine content estimated for the gas phase indicate a non-negligible absorption of excited CF_4. Further investigations are therefore necessary to quantify its contribution and the actual C_3F_8 number density, in particular during the 'on-phase'.

Acknowledgements

This work was supported by the German Research Foundation (DFG) within the framework of the Collaborative Research Centre Transregio 24 ‘Fundamentals of Complex Plasmas’ (SFB-TR 24), projects B2 and B5. The authors thank D Gött, P Druckrey, U Meißner, S Saß, C Senske and F Weichbrodt for valuable technical assistance.

Appendix

If an absorption spectrum of a gas mixture is measured, the absorption from each distinct molecule can be described separately by means of equation (1). In the case of contributions from two molecules (index k = 1 or 2) the total transmitted signal A^j for both considered spectral ranges j = 1 or 2 would be

$$A^1 = A^{11} + A^{12} - A_0^1 \quad \text{(A1)}$$

$$A^2 = A^{21} + A^{22} - A_0^2 \quad \text{(A2)}$$

where A^{jk} denotes the transmitted intensity in the spectral range j for molecule k and $A^j{}_0$ describes the base line for the range j. The number densities n_k can be derived from

$$n_1 = \frac{1}{\sigma_{eff}^{11} \cdot L} \ln\left(\frac{A_0^1}{A^{11}}\right) = \frac{1}{\sigma_{eff}^{21} \cdot L} \ln\left(\frac{A_0^2}{A^{21}}\right) \tag{A3}$$

$$n_2 = \frac{1}{\sigma_{eff}^{12} \cdot L} \ln\left(\frac{A_0^1}{A^{12}}\right) = \frac{1}{\sigma_{eff}^{22} \cdot L} \ln\left(\frac{A_0^2}{A^{22}}\right) \tag{A4}$$

with the effective absorption cross sections σ_{eff}^{jk}. Rearranging equations (A1 - A4) yields a system of two nonlinear equations for two unknown number densities n_1 and n_2

$$\begin{aligned} 1 + \frac{A^1}{A_0^1} &= \exp(-\sigma_{eff}^{11} \cdot L \cdot n_1) + \exp(-\sigma_{eff}^{12} \cdot L \cdot n_2) \\ 1 + \frac{A^2}{A_0^2} &= \exp(-\sigma_{eff}^{21} \cdot L \cdot n_1) + \exp(-\sigma_{eff}^{22} \cdot L \cdot n_2) \end{aligned} \tag{A5}$$

This system can be solved numerically, e.g., with a custom made LabVIEW program in our case, providing for the 4 different and temperature dependent absorption cross sections σ_{eff}^{jk} are known or have been calibrated. Alternatively, (A5) can be rearranged further

$$1 + \frac{A^2}{A_0^2} = \exp(-\sigma_{eff}^{21} \cdot L \cdot n_1) + \left[\left(1 + \frac{A^1}{A_0^1}\right) - \exp(-\sigma_{eff}^{11} \cdot L \cdot n_1)\right]^{\frac{\sigma_{eff}^{22}}{\sigma_{eff}^{12}}} \tag{A6}$$

and is now only a function of n_1 which can be numerically determined while n_2 follows via (A5). Both numerical approaches have been successfully checked. In this context the values I and A (as an integral of I over a spectral micro-window) were used synonymously, since these values are always used relative to the corresponding base line I_0 or A_0, i.e. A can be also considered as an averaged value for I.

References

[1] Gabriel O, Stepanov S and Meichsner J 2007 *J. Phys.* D: *Appl. Phys.* **40** 7383-91
[2] Gabriel O, Stepanov S, Pfafferott M and Meichsner J 2006 *Plasma Sources Sci. Technol.* **15** 858-64
[3] Röpcke J, Mechold L, Käning M, Anders J, Wienhold F G, Nelson D and Zahniser M 2000 *Rev. Sci. Instrum.* **71** 3706-10
[4] Eng R S, Butler J F and Linden K J 1980 *Opt. Eng.* **19** 945-60
[5] Faist J, Capasso F, Sivco D L, Sirtori C, Hutchinson A L and Cho A 1994 *Science* **264**, 553
[6] Beck M, Hofstetter D, Aellen T, Faist J, Oesterle U, Ilegems M, Gini E and Melchior H 2002 *Science* **295** 301
[7] Tuzson B, Mohn J, Zeeman M J, Werner R A, Eugster W, Zahniser M S, Nelson D D, McManus J B and Emmenegger L 2008 *Appl. Phys.* B **92** 451
[8] Hay K G, Wright S, Duxbury G, Langford N 2008 *Appl. Phys.* B **90** 329
[9] Bauer C, Sharma A K, Willer U, Burgmeier J, Braunschweig B, Schade W, Blaser S, Hvozdara L, Müller A and Holl G 2008 *Appl. Phys.* B **92** 327
[10] Cheesman A, Smith J A, Ashfold M N R, Langford N, Wright S and Duxbury G 2006 *J. Phys. Chem.* A **110** 2821
[11] Welzel S, Gatilova L, Röpcke J and Rousseau A 2007 *Plasma Sources Sci. Technol.* **16** 822-31
[12] van Helden J H, Horrocks S J and Ritchie G A D 2008 *Appl. Phys. Lett.* **92** 081506

[13] Stancu G, Lang N, Röpcke J, Reinicke M, Steinbach A and Wege S 2007 *Chem. Vap. Deposition* **13** 351-60
[14] Beyer T, Braun M and Lambrecht A 2003 *Journ. Appl. Phys.* **93** 3158
[15] Röpcke J, Lombardi G, Rousseau A and Davies P B 2006 *Plasma Sources Sci. Technol.* **15** 148
[16] Welzel S, Stepanov S, Meichsner J and Röpcke J 2008 *J.Phys.: Conf. Ser.* submitted
[17] Duxbury G, Langford N, McCulloch M T and Wright S 2005 *Chem. Soc. Rev.* **34** 921
[18] Cruden B A, Rao M V V S, Sharma S P and Meyyappan M 2002 *Plasma Sources Sci. Technol.* **11** 77-90
[19] O'Neill J A, Singh J and Gifford G G 1990 *J. Vac. Sci. Technol.* **A8** 1716

Using quantum cascade lasers with resonant optical cavities as a diagnostic tool

S Welzel[1,5], G Lombardi[2], P Davies[3], R Engeln[4], D Schram[4] and J Röpcke[1]

[1]INP Greifswald, 17489 Greifswald, Felix-Hausdorff-Str. 2, Germany
[2]CNRS LIMHP, Université Paris 13, 99, av. J.B. Clément, 93430 Villetaneuse, France
[3]University of Cambridge, Lensfield Road, Cambridge CB2 1EW, United Kingdom
[4]Eindhoven University of Technology, P.O. Box 513, 5600 MB, Eindhoven, The Netherlands

Email: welzel@inp-greifswald.de

Abstract. Absorption spectroscopy based on optical resonators is known to be a very sensitive diagnostic technique. For many years the infrared spectral range could not be employed, because of the lack of suitable radiation sources with the required power and tunability. Recent advances in semiconductor laser technology, in particular the advent of quantum cascade lasers (QCL) provides new possibilities for highly sensitive and selective detection of molecular species. Cavity Enhanced Absorption Spectroscopy (CEAS) with a thermoelectrically (TE) cooled cw QCL emitting at 7.66 μm and employing a ~ 0.5 m long cavity yielded effective path lengths of 1080 m and a sensitivity of 2×10^{-7} $cm^{-1}Hz^{-1/2}$, mainly limited by incomplete averaging over cavity resonances. The molecular concentration detection limit with a 20 s integration time was found to be 6×10^8 molecules/cm^3 for N_2O and 2×10^9 molecules/cm^3 for CH_4 which is good enough for the selective measurement of trace atmospheric constituents at 2.2 mbar.

1. Introduction

Absorption spectroscopy in the mid infrared spectral region (MIR, 3 - 20 μm) using lasers as radiation sources is an effective method for monitoring molecular species in the fingerprint region. Conventional linear absorption spectroscopy employing multiple pass cells is a well-known method for achieving high sensitivity [1,2]. An alternative approach where path lengths up to several kilometres can be achieved is to combine a high finesse optical resonator with a pulsed or continuous wave (cw) source respectively for cavity ring-down spectroscopy (CRDS) [3], cavity enhanced absorption spectroscopy (CEAS) [4], or integrated cavity output spectroscopy (ICOS) [5,6]. In CEAS or ICOS the steady-state transmission or integrated transmitted intensity through the cavity is observed as a function of the laser frequency.

The majority of cavity based methods have used sources of radiation in the ultraviolet and visible regions. For many years the infrared spectral range could not be employed either for CRDS or for the CEAS or ICOS techniques, because of the lack of suitable radiation sources with the required power and tunability. Early experiments in the MIR were therefore carried out with bulky radiation sources,

[5] To whom any correspondence should be addressed.

e.g. optical parametric oscillators [7,8], Raman cells or shifters [9-11] or CO_2 and CO lasers [12-14]. In all these cases sophisticated optical geometries were developed which were more suitable for the research laboratory than for field applications. The combination of lead salt lasers with optical cavities suffered from low laser intensity [15]. Recent advances in semiconductor laser technology, in particular the advent of intersubband quantum cascade lasers (QCL) and interband cascade lasers (ICL) [16,17], provides new possibilities for highly sensitive and selective trace gas detection. Sensitivities of 5×10^{-10} $cm^{-1}Hz^{-1/2}$ were accomplished by means of multiple pass cells but at the expense of large sample volumes of a few litres and increased base lengths [2]. Distributed feedback (DFB)-QCLs combine single-frequency operation with tunability over several wavenumbers, and average powers over a mW. The relatively high output power permits the use of optical cavities with high finesse. The potential of such a combination was demonstrated in CRDS measurements with a cryogenically cooled cw DFB-QCL [18]. Several applications of QCLs combined with either the CRDS or the CEAS/ICOS techniques and with sensitivities down to 2×10^{-9} $cm^{-1}Hz^{-1/2}$ were reported [19]. The broader use of cavity based techniques in the MIR in industry or for field application has so far been hindered not only by space or weight restrictions, but also by the need for cryogenic cooling of either the laser or the detector. While pulsed QCLs working at room temperature have been commercially available for several years, room-temperature cw QCLs have only recently been introduced.

Moreover pulsed QCLs exhibit an intrinsic frequency chirp appearing as an increased effective laser line width [20] and exceeding the width of narrow molecular absorption features. Thus the application of TE cooled pulsed QCLs for ICOS and CRDS was found to be limited by the inevitable excitation of multiple cavity modes during one pulse [21-24]. Consequently the accuracy and advantage of an absolute internal absorption calibration, in theory inherent for CRDS experiments, is not achievable. Additionally, the chirped laser does not allow for an efficient build-up of the laser field in the cavity. CRDS using pulsed QCLs therefore has only a limited number of useful applications, e.g. for the determination of the reflectivity of the cavity mirrors without any absorption feature present or for the detection of complex and broad molecular absorptions at higher pressures where the width of absorption features exceeds the effective laser line width.

The essential narrow bandwidth emission from a QCL in combination with an optical resonator can be gained if the current of the laser is not switched or pulsed. Therefore a room temperature cw QCL emitting at 7.66 μm was used in a CEAS configuration aimed at achieving a sensitive, cryogen-free spectrometer for sampling small volumes. Apart from several studies on NO at 5 μm the MIR spectral region in general and the 7 - 8 μm region presently used have not been extensively investigated. The experimentally straightforward CEAS configuration would be of special interest for *in-situ* measurements of radicals, e.g. in low pressure plasmas, where multi-pass cell sampling *ex-situ* is not an option and where CRDS with a pulsed QCL suffers from the laser chirp and the low pressure, i.e. the width of the absorption lines cannot be arbitrarily matched to the effective pulsed QCL line width.

2. Experimental

The experimental arrangement for the CEAS experiments is shown in figure 1. The stable resonator was formed by two high reflectivity mirrors (Los Gatos Research) of diameter 25.4 mm and 1 m radius of curvature. It should be noted that a high reflectivity of the mirrors of better than 99 % can only be achieved for a relatively small wavelength window of typically 150 cm^{-1}. The mirrors also served to enclose the vacuum vessel which was made of standard vacuum components. Hence the vacuum cell determined the mirror separation of 0.432 m. In contrast to locked cavity studies, the cavity length here was not actively changed or dithered nor was the cavity locked to the wavelength of the laser source. Beam shaping optics (BSO) collected the radiation from the QCL firstly with an aspherical ZnSe lens (30 mm diameter, f/1.5) and then with a telescope formed by two off-axis parabolic mirrors (OAP) which reduced the beam diameter by a ratio of almost 1:5. No additional optics were added to suppress possible optical feedback to the laser, since feedback was assumed to be negligible for the current DFB-QCL device. The radiation leaking out of the cavity was collected by an OAP of 25.4 mm diameter (f/0.64) and directed to a detector. A reference path could be used for calibration by inserting a beam flipping mirror. In the beginning, both signal and reference channel were equipped with liquid nitrogen (LN) cooled detectors (Judson Technologies). In addition, for validation experiments a TE cooled detector (neoplas control, VIGO element PDI 3TE 10/12) was employed for the signal channel.

A room temperature cw QCL (Alpes Lasers), emitting between 1300 cm^{-1} to 1311 cm^{-1} (~ 7.66 μm) was used for the experiments. It was operated in a standard housing (Starter Kit, Alpes Lasers), thermoelectrically cooled to - 5.9°C for calibration or + 2.2°C for trace gas measurements in room air and driven by a DC current source (Kepco, BOP100-2M) with a typical DC current of around 380 mA. An additional current ramp of 45 mA was impressed on it in order to sweep the laser frequency during the on-phase. The laser current was briefly reduced below the threshold value by -70 mA to record the offset of the zero signal (figure 2) in the off-phase. The sweep rate of the laser was 0.9 kHz for all measurements. For a mirror distance of 0.432 m the sweep of the laser frequency corresponded to a scan over 67 cavity modes, which was sufficient to avoid an overlap between the build-up and decay events of different cavity modes. On the other hand the sweep rate gave a full spectrum after slightly more than 1 ms. A given number of spectra were then averaged and simultaneously fitted. In order to smooth interference effects on the baseline of the spectra, mainly caused by longitudinal cavity modes, the whole cavity was arbitrarily destabilized mechanically, e.g. by an external vibrating source.

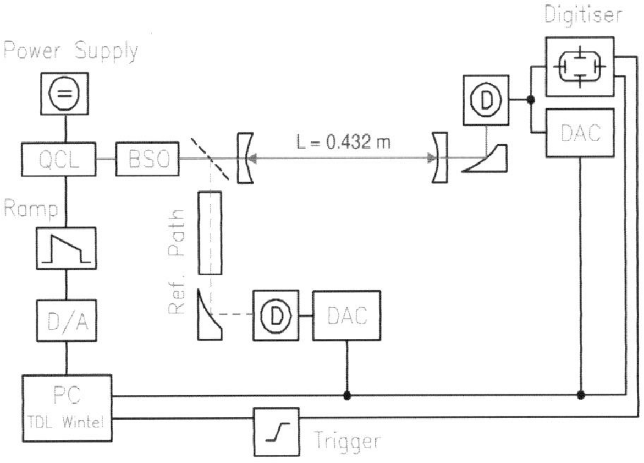

Figure 1. Schematic diagram of the apparatus. Beam shaping optics (BSO) provided efficient light transmission to the resonator. The detector (D) signals were recorded with data acquisition cards (DAC). The QCL was driven by a cw source and the tuning ramp for the laser was generated via the digital-to-analog-converter (D/A).

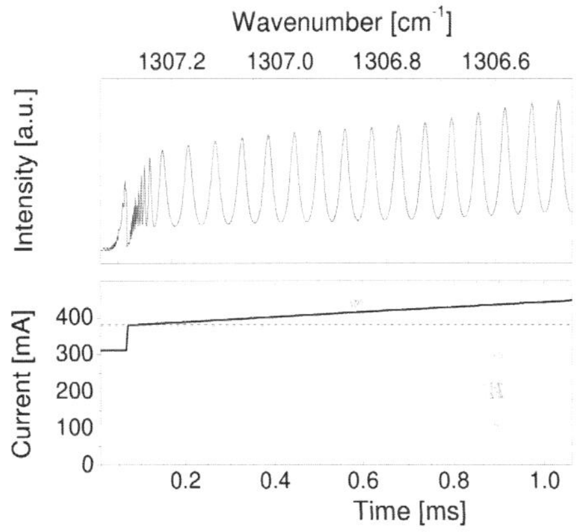

Figure 2. Lower: Average current (dashed line) and its modulation (heavy black line) applied to the cw QCL during a single sweep. Upper: Signal through a 25.4 mm Germanium etalon demonstrating the 0.78 cm^{-1} sweep of the laser when the modulation ramp was applied.

The laser sweep of 0.78 cm^{-1} or 23.4 GHz, with the current modulation applied, is illustrated at the top of figure 2 which shows the signal through a Germanium etalon (0.048 cm^{-1} free spectral range). Relative calibration of the frequency axis of the spectrum was provided by recording the fringes of a Germanium etalon in a reference channel and using interpolation. Absolute calibration and identification was achieved by comparison with standard gas absorption spectra recorded in the reference channel prior to the CEAS spectra.

3. Results

3.1. Calibration of the CEAS system

Spectra from different mixtures of N_2O in N_2 were recorded in order to determine the effective reflectivity of the cavity mirrors and to provide a calibration. The measurements were taken under flowing conditions at a constant pressure of 1.2 mbar. Figure 3 shows the result of the spectral scans, obtained with a liquid nitrogen cooled detector and averaged over 20 s to reduce the noise level below ~ 2×10^{-3} in transmission. The three dominant N_2O features appear at 1306.771 cm^{-1}, 1306.846 cm^{-1} and 1306.929 cm^{-1} respectively. The strongest line has a line strength S of 1.0×10^{-19} cm/molecule. For the lower intensity lines S values of 9.7×10^{-21} cm/molecule are reported in the literature [25].

The absorption coefficient $k(\nu)$, i.e. the product of the number density n and the absorption cross section σ, is inferred from CEAS spectra through

$$n\sigma(\nu) = k(\nu) = \left(\frac{I_0(\nu)}{I(\nu)} - 1\right)\frac{1-R}{L} \tag{1}$$

where L is the mirror spacing, I_0 and I are the transmitted intensity of an empty cavity (baseline) and the cavity filled with the species of interest respectively. Obviously the reflectivity R of the cavity mirrors must be known or calibrated in order to deduce absolute values for k. A fit of the transmission spectra (figure 2) resulted in an effective reflectivity of $R = 99.96$ % for the cavity mirrors or an effective path length $L_{eff} = L/(1-R)$ of 1080 m. Since in this approach the excitation of high order transverse modes may occur which could have different diffraction losses this value of R should be regarded as an effective one [26]. No evidence was found that optical feedback from the cavity to the QCL caused line shifts or perturbed lines. Next, the three N_2O lines visible in figure 3 were analysed further using

$$\ln\left(\frac{I_0}{I}\right) \approx GA \sim n \tag{2}$$

where $G = R/(1 - R)$ and $A = (1 - \exp(-kL))$ [19]. If the natural logarithm is taken from the inverted transmission spectrum and plotted against n a linear relationship is expected, as long as a weak absorber is present in the cavity.

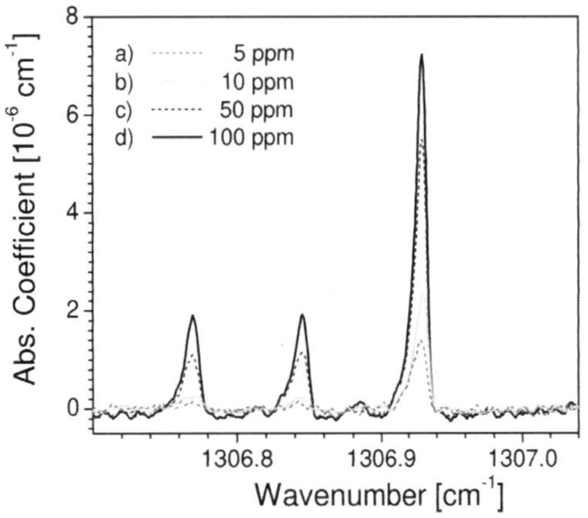

Figure 3. CEAS scans for different N_2O calibration mixtures (N_2O diluted in N_2) at a constant total pressure of 1.2 mbar: a) 5 ppm, b) 10 ppm, c) 50 ppm, d) 100 ppm The acquisition time was 20 s.

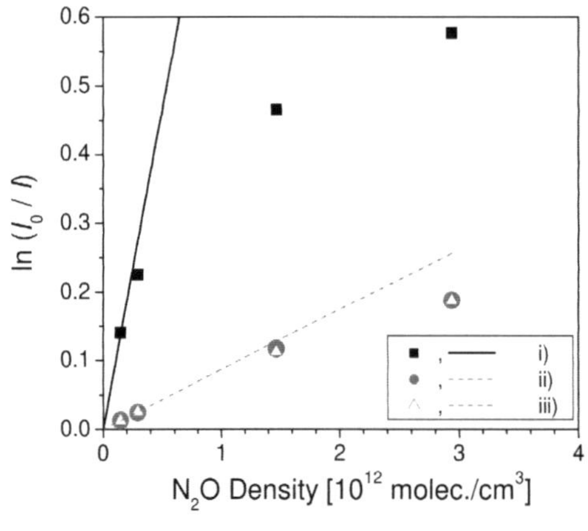

Figure 4. Calibration data for the maxima of the three absorption lines and the four mixing ratios of N_2O mentioned in figure 3: i) 1306.929 cm^{-1}, ii) 1306.846 cm^{-1} and iii) 1306.771 cm^{-1}.

Thus equation (2) serves as a check on the validity of the weak absorption limit ($k \rightarrow 0$ and $R \rightarrow 1$). For this purpose $\ln(I_0/I)$ at the maximum of the N_2O lines was plotted against the concentration in the cavity (figure 4). Generally, it can be concluded that the linear approximation is valid for GA ≤ 0.15 whilst for absorption features bigger than 15 % a correction is necessary.

3.2. Test of validity and system performance

The CEAS system was validated with measurements on constituents of laboratory air, namely methane, nitrous oxide and water. In order to detect these species within one spectral scan the spectral range and therefore the heat sink temperature of the QCL was slightly changed compared to the calibration experiments. The spectrum expected to arise for the average mixing ratios of the three molecules was calculated and is plotted in the upper panel of figure 5. Some non interfering absorption features were selected for analysis and these are summarized in table 1. The single water line in the spectrum consists of two unresolved lines with a spacing of 0.004 cm^{-1}, so that an effective value for S (for simplicity the sum) is used whereas a third weaker line shifted by ~ 0.01 cm^{-1} could be deconvoluted from the main lines and was not used for the quantitative analysis. The transmission spectrum recorded with a TE cooled detector is shown in the lower panel of figure 5. The measurements were performed at low pressure under flowing conditions at 2.2 mbar in order to increase line selectivity. The spectrum was averaged for 20 s. Qualitatively it agrees very well with the simulation. The line positions were fitted to a transmission profile (figure 5). This fit was then used to calculate the absorption coefficient using (1). Integrating over the absorption line of interest revealed the number densities and mixing ratios (table 1): 1.7 ppm of CH_4, 350 ppb N_2O and more than 1 % of H_2O. A comparison with literature values for the abundance of these molecules in air (1.75 ppm for CH_4, 320 ppb for N_2O and 0.1 ... 4 % for H_2O) yields fair agreement [27,28].

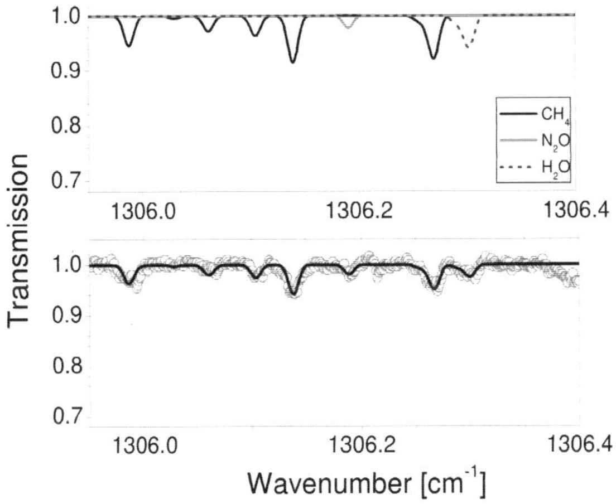

Figure 5. Calculated spectrum (upper) for air containing CH_4, N_2O and H_2O. Experimental conditions: 2.2 mbar, L_{eff} = 1080 m. The corresponding CEAS spectrum (lower) was observed with a TE cooled detector (open circles) and fitted (solid line) to determine the actual concentrations.

The measurements on laboratory air were repeated with a LN cooled detector and the results, in particular the relative measurement error after 20 s, were very similar. For the LN cooled detector $\Delta I/I_0$ of 0.3 % was obtained whereas for the TE cooled detector 0.2 % was found. The corresponding noise equivalent absorption (NEA), defined by

$$\mathrm{NEA} = \frac{\Delta I}{I_0}\left(\frac{1-R}{L}\right)\sqrt{\frac{2}{f}} \tag{3}$$

Where f is the acquisition rate of the spectra, is then 1.8×10^{-7} $cm^{-1}Hz^{-1/2}$ for the TE cooled detector if the calibration error in R is also included. It is possible to estimate the minimum detectable number density (MDND) n_{min} for each molecule by means of the absorption coefficient k_{min}. Assuming that k_{min} corresponds to the maximum of the smallest absorption feature that can be distinguished from the noise, the n_{min} can be estimated. The MDNDs of the different molecules analysed are listed in table 1. For N_2O a detection limit of 6.4×10^8 molecules/cm^3 is obtained, which transferred to CH_4 corresponds to 1.6×10^9 molecules/cm^3 for the strongest line in CH_4. If this is converted into a mixing ratio at 2.2 mbar the limits of detection for N_2O and CH_4 are 12 ppb and 46 ppb respectively.

Table 1. Species detected in ambient laboratory air at 2.2 mbar (n - number density).

Molecule	ν [cm^{-1}]	Line strength S [cm/molecule]	Mix. Ratio	n [molecules/cm^3]	MDND (n_{min}) [molecules/cm^3]
CH_4	1305.987	4.78×10^{-20}	1.72 ppm	9.2×10^{10}	2.5×10^9
CH_4	1306.062	2.41×10^{-20}	1.82 ppm	9.8×10^{10}	4.9×10^9
CH_4	1306.105	3.19×10^{-20}	1.68 ppm	9.0×10^{10}	3.7×10^9
CH_4	1306.140	7.60×10^{-20}	1.66 ppm	8.9×10^{10}	1.6×10^9
N_2O	1306.191	1.12×10^{-19}	346 ppb	1.9×10^{10}	6.4×10^8
H_2O	1306.29	6.08×10^{-24} [a)]	1.1%	5.8×10^{14}	1.8×10^{13}

[a)]effective line strength from two unresolved lines

4. Discussion and Summary

A thermoelectrically cooled QCL has been combined with a high finesse optical resonator in order to profit from the enlarged path lengths at reasonably small sample volumes in combination with the high absorption cross section in the infrared molecular fingerprint region. The sensitivity of 1.8×10^{-7} cm^{-1}Hz$^{-1/2}$ presently achieved with the TE cooled CEAS system is comparable to other CEAS or ICOS studies exhibiting a NEA down to 4×10^{-8} cm^{-1}Hz$^{-1/2}$ [26,29[6]]. In comparison to entirely TE cooled based CRDS setups with pulsed QCLs the performance of the system is superior. For CRDS employing pulsed QCLs the fast frequency chirp of the laser causes an excitation from several to tens of cavity modes almost simultaneously. The result is an apparent linewidth effects, i.e. line broadening, loss of sensitivity and accuracy. It is concluded that the switching of the laser current and hence the rapid temperature changes of the QCL device should be avoided in order to benefit from the combination with a resonant cavity [21]. Narrow bandwidth radiation of typically less than 100 MHz (≤ 0.003 cm^{-1}) can be provided by cw QCLs which is sufficient for the combination with the presently used cavity of 0.432 m and its free spectral range of 360 MHz (0.012 cm^{-1}).

The NEA of the system is still almost a factor of ~ 50 less sensitive than CRDS employing a cw QCL as demonstrated by Paldus et al. [18] which may be explained by the commonly omitted cavity mode noise in CRDS. In the present case the system was aligned on-axis with respect to the laser beam which produces cavity mode noise inherent to CEAS/ICOS setups [29]. In order to achieve an accurate signal-to-noise ratio, a measurement interval of at least 5 s is necessary to smooth out the residual mode structure on the baseline. For CEAS or ICOS the minimum detectable absorption is typically limited to 10^{-2} to 10^{-3} due to incomplete averaging over the cavity resonances. Since this still dominates the total performance the choice of detector, i.e. TE or LN cooled, is of less importance for the sensitivity limits of the system as demonstrated in section 3. Due to an advantageous spectral position and a lower sampling pressure in our experiment the MDND of CH_4 was lower than in similar measurements carried out at 7.9 µm where a better sensitivity of 2×10^{-9} cm^{-1}Hz$^{-1/2}$, but a $n_{min}(CH_4)$ of 2.9×10^9 molecules/cm^3 were reported [19]. With a 20 s measurement interval detection limits for N_2O and CH_4 of 6×10^8 molecules/cm^3 and 2×10^9 molecules/cm^3 respectively could be achieved at 2.2 mbar indicating that sub-ppb levels could easily be measured at higher pressures.

Mode noise is absent in QCL spectrometers employing conventional long path cells which are today increasingly based on TE cooled lasers and detectors and hence exhibit higher sensitivities even down to 3×10^{-10} cm^{-1}Hz$^{-1/2}$ [30]. However, the volume of commercial multiple pass cells covering effective path lengths up to 210 m, typically ranges from 0.5 to 5 l which may increase the dimensions of the system

[6] Value estimated by the authors from the cited data.

and the necessary pumping time. The aim has been to achieve maintenance free operation with a small sample volume, presently around 0.3 l, and with an optical geometry reduced in its complexity which would make the system field-deployable. The results achieved here are relevant both for the *in-situ* detection of radicals in plasma chemistry and for trace gas measurements with field-deployable systems. Additionally, trace gas measurements may also be carried out without pre-concentration procedures. The choice of an appropriate method depends on whether the important criterion is ultimate sensitivity or a more compact system. In the former case a multi pass cell spectrometer would be preferable because of its better signal to noise characteristics but this configuration would exclude the detection of processes on short time scales or certain types of *in-situ* measurements. To achieve a compact system small volume cavity based spectrometers employing cw QCLs would be more appropriate.

Acknowledgement
This work was supported by the German Research Foundation (DFG) within the framework of the Collaborative Research Centre Transregio 24 "Fundamentals of Complex Plasmas" (SFB-TR 24). K.-D. Weltmann is gratefully acknowledged for his continuous encouragement and general support making the present work possible.

References

[1] Röpcke J, Lombardi G, Rousseau A, and Davies P B 2006 *Plasma Sources Sci. Technol.* **15** S148
[2] Nelson D D, Shorter J H, McManus J B, and Zahniser M S 2002 *Appl. Phys.* B **75** 343
[3] O'Keefe A and Deacon D A G 1988 *Rev. Sci. Instrum.* **59** 2544
[4] Engeln R, Berden G, Peeters R, and Meijer G 1998 *Rev. Sci. Instrum.* **69** 3763
[5] O'Keefe A, Scherer J J, and Paul J B 1999 *Chem. Phys. Lett.* **307** 343
[6] O'Keefe A 1998 *Chem. Phys. Lett.* **293** 331
[7] Scherer J J, Voelkel D, Rakestraw D J, Paul J B, Collier C P, Saykally R J, and O'Keefe A 1995 *Chem. Phys. Lett.* **245** 273
[8] Scherer J J, Aniolek K W, Cernansky N P, and Rakestraw D J 1997 *J. Chem. Phys.* **107** 6196
[9] Ito F 2006 *Journ. Chem. Phys.* **124** 054309
[10] Paul J B and Saykally R J 1997 *Anal. Chemistry* **69** 287 A
[11] Paul J B, Provencal R A, Chapo C, Roth K, Casaes R, and Saykally R J 1999 *J. Phys. Chem.* A **103** 2972
[12] Mürtz M, Freech B, and Urban W 1999 *Appl. Phys.* B **68** 243
[13] Peeters R, Berden G, Olafsson A, Laarhoven L, and Meijer G 2001 *Chem. Phys. Lett.* **337** 231
[14] Halmer D, von Basum G, Hering P, and Mürtz M 2005 *Opt. Lett.* **30** 2314
[15] Remy J, Hemerik M M, Kroesen G M W, and Stoffels W W 2004 *IEEE Trans. on Plasma Science* **32** 709
[16] Faist J, Capasso F, Sivco D L, Sirtori C, Hutchinson A L, and Cho A 1994 *Science* **264** 553
[17] Yang R Q, Hill C J, Yang B H, Wong C M, Muller R E, and Echternach P M 2004 *Appl. Phys. Lett.* **84** 3699
[18] Paldus B A, Harb C C, Spence T G, Zare R N, Gmachl C, Capasso F, Sivco D L, Baillargeon J N, Hutchinson A L, and Cho A Y 2000 *Opt. Lett.* **25** 666
[19] Paul J B, Scherer J J, O'Keefe A, Lapson L, Anderson J G, Gmachl C, Capasso F, and Cho A Y 2002 *Proc. SPIE* **4577** 1
[20] Duxbury G, Langford N, McCulloch M T, and Wright S 2005 *Chem. Soc. Rev.* **34** 921
[21] Welzel S, Lombardi G, Davies P B, Engeln R, Schram D C, and Röpcke J 2008 *Journ. Appl. Phys.* **104** *in press*
[22] Silva M L, Sonnenfroh D M, Rosen D I, Allen M G, and O'Keefe A 2005 *Appl. Phys.* B **81** 705
[23] Manne J, Sukhorukov O, Jäger W, and Tulip J 2006 *Appl. Opt.* **45** 9230
[24] Sukhorukov O, Lytkine A, Manne J, Tulip J, and Jäger W 2006 *Proc. SPIE* **6127** 61270A
[25] Rothman L S et al. 2005 *Journ. Quant. Spectrosc. Radiat. Transf.* **96** 139
[26] Menzel L, Kosterev A A, Curl R F, Tittel F K, Gmachl C, Capasso F, Sivco D L, Baillargon J N, Hutchinson A L, Cho A Y, and Urban W 2001 *Appl. Phys.* B **72** 859

Journal of Physics: Conference Series **157** (2009) 012009 doi:10.1088/1742-6596/157/1/012009

[27] Brown S S 2003 *Chem. Rev.* **103** 5219
[28] Uherek E and Sander R 2006 *ESPERE Climate Encyclopaedia* (http://www.espere.net/)
[29] McCurdy M R, Bakhirkin Y A, and Tittel F K 2006 *Appl. Phys.* B **85** 445
[30] Nelson D D, McManus J B, Herndon S C, Shorter J H, Zahniser M S, Blaser S, Hvozdara L, Muller A, Giovannini M, and Faist J 2006 *Opt. Lett.* **31** 2012

Journal of Physics: Conference Series **157** (2009) 012010 doi:10.1088/1742-6596/157/1/012010

Application of quantum cascade laser absorption spectroscopy to studies of fluorocarbon molecules

S Welzel[1,3], S Stepanov[2], J Meichsner[2] and J Röpcke[1]

[1]Leibniz Institute for Plasma Science and Technology, Felix-Hausdorff-Str. 2, D-17489 Greifswald, Germany
[2]Institute of Physics, Ernst-Moritz-Arndt University, Felix-Hausdorff-Str. 6, D-17489 Greifswald, Germany

Email: welzel@inp-greifswald.de

Abstract. The recent advent of quantum cascade lasers (QCLs) enables room-temperature mid-infrared spectrometer operation which is particularly favourable for industrial process monitoring and control, i.e. the detection of transient and stable molecular species. Conversely, fluorocarbon containing radio-frequency discharges are of special interest for plasma etching and deposition as well as for fundamental studies on gas phase and plasma surface reactions. The application of QCL absorption spectroscopy to such low pressure plasmas is typically hampered by non-linear effects connected with the pulsed mode of the lasers. Nevertheless, adequate calibration can eliminate such effects, especially in the case of complex spectra where single line parameters are not available. In order to facilitate measurements in fluorocarbon plasmas, studies on complex spectra of CF_4 and C_3F_8 at 7.86 µm (1269 - 1275 cm^{-1}) under low pressure conditions have been performed. The *intra-pulse* mode, i.e. pulses of up to 300 ns, was applied yielding highly resolved spectral scans of ~ 1 cm^{-1} coverage. Effective absorption cross sections were determined and their temperature dependence was studied in the relevant range up to 400 K and found to be non-negligible.

1. Introduction

Mid-infrared (IR) laser absorption spectroscopy (AS) has been known as a versatile tool for the detection of molecular species in their fingerprint region (3 - 20 µm) for years, whereas detection schemes employing interband and quantum cascade lasers (QCLs) have only recently been introduced owing to the increasing availability of this new type of lasers. QCLAS in the gas phase is now applied in various fields covering environmental trace gas detection [1-3], detection of toxic species or explosives [4,5], isotope ratio measurements [6], breath gas analysis [7,8] as well as industrial process monitoring [9] and plasma diagnostics [10-12]. The first demonstration of a QCL in 1994 [13], based on the idea of Kazarinov and Suris [14], and further development enable now pulsed or continuous wave, thermoelectrically (TE) cooled, single-mode distributed feedback (DFB-)QCLs to be used in spectrometers.

Different modes can be used to tune a QCL over the spectral transitions of interest. Particularly the *intra-pulse* mode is of high interest for time resolved measurements. For this tuning approach a relatively long laser pulse is used. The internal heating of the QCL causes a frequency down-chirped

[3] To whom any correspondence should be addressed.

laser pulse which can be used to acquire a full absorption spectrum within the pulse width of typically a few hundred nanoseconds [15]. However, at low pressure conditions the measured absorption features are affected by an obstacle known as the "rapid passage effect" accompanied by potential power saturation effects [16,17]. Nevertheless, if a sensible normalisation [12] or calibration [9,11] is performed quantitative results can be obtained. This also addresses the required high resolution spectra of molecules with a complex structure which are typically absent or difficult to measure in jet cooled experiments. In this case a calibrated effective absorption cross section determined for a distinct spectral micro-window is also sufficient: Stancu et al. demonstrated this for SiF_4 where a temperature dependence was not observed, because the different contributions from several spectral lines tended to compensate each other [9].

In combination with TE cooled detectors QCL spectrometers became cryogen free and are hence particularly favourable for industrial applications and process monitoring. Moreover, the achievable time resolution with pulsed QCLs is theoretically limited to their pulse width on the nanosecond time scale and thus interesting for basic research on plasma chemical phenomena [11]. The QCL based detection and monitoring of stable and transient molecules in pulsed discharges is therefore an issue in both fields. Studies on pulsed discharges are essential for a better understanding of the plasma chemistry in usually complex industrial discharges such as fluorocarbon containing gas mixtures. IR tunable diode laser absorption spectroscopy in general provides an adequate means for measurements in these discharges [18-21] and QCLAS may facilitate a better time resolution.

Since pulsed fluorocarbon plasmas are of high interest in industry and as a model system for plasma chemical investigations the present study is concerned about the detection of stable fluorocarbons, among them the precursor CF_4. Measurements were carried out in pulsed radio-frequency (rf) discharges at 7.86 µm (1269 - 1275 cm^{-1}) by means of a pulsed QCL. The experimental arrangement is discussed in section 2. Both fluorocarbons, CF_4 and C_3F_8, with absorption features in this spectral range exhibit complex spectra [22-25]. For C_3F_8 only coarse spectra or band integrated absorption cross sections are available in literature. For CF_4 a few IR tunable diode laser measurements, i.e. highly resolved spectroscopic data, around the band origin at 1283 cm^{-1} exist, but these studies provide only line positions [26,27] or relative absorption cross sections [28]. Measurements around the band origin of CF_4 were presently not feasible because of the almost total absorption of the IR radiation by the precursor molecule under the present experimental conditions. For that reason effective absorption cross sections have been determined for both molecules which is described in detail section 3. This comprises also the definition of two different spectral ranges for the plasma measurements in order to facilitate a discrimination between the contributions from the different species. Furthermore the temperature dependence of the absorption cross sections has been studied which is also discussed in section 3. The results are summarised in section 4.

2. Experimental

The focus of the present study was twofold: (i) the design of a IR spectrometer for time resolved QCLAS measurements in pulsed radio-frequency (rf) discharges [29] and (ii) the determination of necessary calibration data for the absorption features of CF_4 and C_3F_8. Although not necessary in terms of a high time resolution or special trigger events all calibration data have been recorded under identical trigger conditions as for the plasma measurements. This approach was required, since changes in the trigger frequency (e.g., to speed up the data acquisition) also affect the heat sink temperature and hence the output frequency of the QCL.

The entire optical set-up including triggering devices is shown in figure 1. The QCL optics was located on an optical board (OB I) and provided a collimated IR beam which was co-aligned to an existing tunable diode laser arrangement (OB II) [21]. A part of this beam shaping optics was used to guide the QCL beam in double pass configuration (~ 90 cm) through the plasma reactor. A fast TE cooled detector (D_2, neoplas control, VIGO element PDI 3TE-10/12) was placed close to the beam exit at the plasma chamber where the transmitted light was collected with an off-axis parabolic mirror (OAP) (25.4 mm diameter, f/0.64). First, the emitted QCL radiation was collected with an OAP

(50 mm diameter, f/0.99). Next, in order to transmit a maximum of the laser output power through the shared ~ 30 mm entrance and exit port of the reactor, the beam diameter was reduced by means of an OAP telescope by a factor of almost 4.

For further analysis of QCL spectra acquired during a chirped laser pulse reliable triggering is essential because a single absorption line typically spreads over less than 5 ns. A jitter of ~ 1 ns between the electrical trigger of the power supply and the detected laser signal induces thus a non-negligible error on the results. In order to tackle this jitter a second fast TE cooled detector (D_1) has been installed in a reference path. The rising edge of the laser pulse deflected by a ZnSe beam splitter onto this detector provided a stable trigger for data acquisition with a fast digitising oscilloscope (LeCroy, WR104Xi).

The pulsed room-temperature DFB-QCL (Alpes Lasers) emitting between 1269 and 1275 cm^{-1} (7.86 μm) was operated with a QCL measurement and control system (Q-MACS, neoplas control) in a water cooled laser housing. All measurements were carried out in the *intra-pulse* mode with a laser pulse width of 300 ns, i.e. a spectral coverage of ~ 1 cm^{-1}, and repetition frequencies between 200 Hz and 1 kHz. The trigger to the QCL power supply was provided by a frequency generator (Hewlett Packard) activated by a master pulse generator (MPG, BNG 555) which normally also served as the trigger for the rf discharge. More details on the trigger scheme can be found in [29]. The measurements were performed at two spectral positions around 1271 cm^{-1} and 1274 cm^{-1} requiring a QCL heat sink temperature of + 17 °C and - 15 °C respectively. Relative and absolute calibration of the laser frequency was provided by the fringes of a germanium etalon (25.4 mm, 0.0485 cm^{-1} free spectral range) and interpolation as well as the spectral features of CH_4 and N_2O.

In order to determine effective absorption cross sections σ_{eff} for CF_4 and C_3F_8 corresponding gas mixtures were measured in the plasma reactor. However, the temperature dependence $\sigma_{eff}(T)$ could not be deduced with this procedure, because a homogeneous external heating of the gas phase in the reactor volume of 20 l was not feasible. Therefore a 25 cm long metallic reference cell wrapped with a heating tape was filled with the studied fluorocarbons and placed inside the beam path (figure 1).

3. Results and discussion

3.1. Basic spectroscopic issues

For measurements in the discharge a spectral position with reduced interferences from other plasma produced transient or stable fluorocarbons was required. Figure 2 (upper panel) shows normalised FTIR spectra for three stable C_xF_y molecules in the considered spectral range. Taking into account the fact that the precursor CF_4 could not be measured at the maximum of its absorption band, since the transmission through the reactor would have been too weak in this case, the spectral window between 1269 and 1275 cm^{-1} has been chosen. Other stable molecules than C_3F_8 may not interfere in this target range. The CF_3 and CF radicals (figure 2, lower panel) would only give minor contributions due to weak absorption lines or low concentrations of CF.

Highly resolved absorbance spectra of CF_4 and C_3F_8 measured in the considered spectral range by means of QCLAS are shown in figure 3. These overview scans consist of several successively acquired parts obtained by tuning the heat sink temperature of the QCL while keeping pulse width and repetition frequency constant. C_2F_6 was also checked but exhibited no considerably absorption in this range. While the C_3F_8 absorbance appears to be smooth over the mapped frequency range strong fluctuations on the CF_4 data are observed. As shown in the inset of figure 3, in the case of CF_4 single lines or a group of lines can still be resolved. In combination with the ringing features of the "rapid passage effect" this spectrum appears as "noisy" whereas the C_3F_8 spectrum is very dense and the mentioned effects smooth out.

Furthermore, two spectral micro-windows (R_1, R_2) around 1271 cm^{-1} and 1274 cm^{-1}, respectively, are marked in figure 3. As can be clearly seen from the graph the ratio of the absorption cross sections of both stable molecules is reversed for R_1 and R_2. Thus, if absorption spectra are acquired in these micro-windows under identical discharge conditions this provides a means to distinguish between CF_4

and C_3F_8 contributions and to derive their concentrations. Details on this approach can be found in [29]. In what follows the effective absorption cross sections are calibrated for R_1 and R_2.

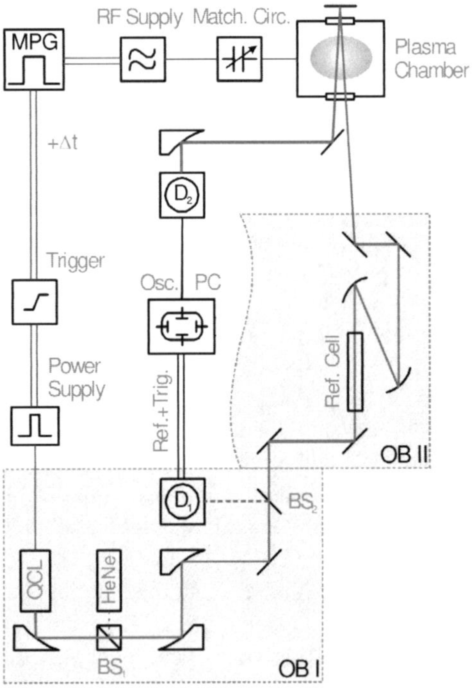

Figure 1: Schematic diagram of the apparatus used for calibration measurements and subsequent plasma diagnostic studies (cf. [29]). Double lines represent trigger lines. (OB - optical board, BS - beam splitter, D - detector, MPG - master pulse generator, Osc. - oscilloscope)

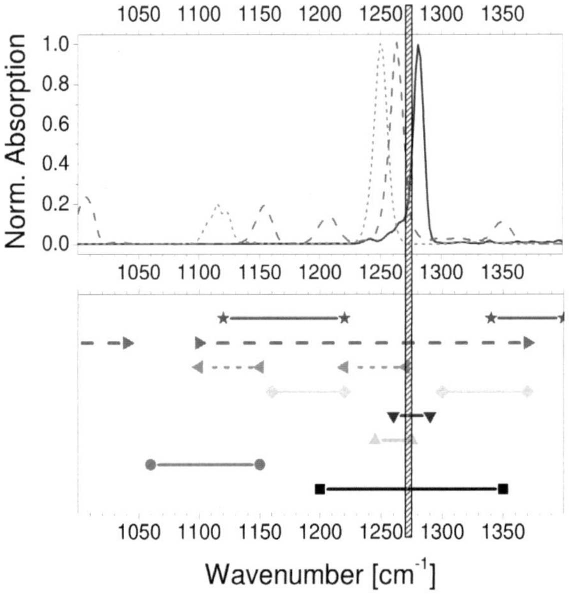

Figure 2: Spectral ranges with absorption features of different fluorocarbon molecules and selected spectral window for the present study (shaded area). Upper: Normalised FTIR spectra of CF_4 (solid line), C_3F_8 (dashed line) and C_2F_6 (dotted line). Lower: Potential interferences of C_xF_y absoprtion bands (■ - CF, ● - CF_2, ▲ - CF_3, ▼ - CF_4, ◆ - C_2F_4, ◀ - C_2F_6, ▶ - C_3F_8, ★ - CHF_3) [22,25,30,31]

The recorded absorption spectra were affected by three main noise sources: i) noise from the preamplifier of the detector, ii) QCL pulse to pulse intensity fluctuations, and iii) noise induced from the RF generator during plasma measurements, particularly, at the transition points of the pulsed discharge. Averaging over 25 QCL pulses provided a sufficiently smoothed signal for further analysis. For calibration measurements, however, averaging was performed over 600 pulses, mainly to keep the trigger scheme (i.e. the duty cycle) and thus the heat sink temperature and spectral output of the QCL stable. An example for several overlaid single shot spectra and an averaged spectrum (over 600 pulses) is given in figure 4.

3.2. Determination of effective absorption cross sections

In absorption spectroscopy the number density n is inferred from Beer-Lambert's law for a single line from the ratio of the incident and transmitted intensity I_0 and I in a sample with an absorption length L

$$\ln(I_0(\nu)/I(\nu)) = n\sigma(\nu)L \tag{1}$$

where $\sigma(\nu)$ is the absorption cross section. Equation (1) is often used in its integral form over the transition of interest where the frequency dependence of $\sigma(\nu)$ is replaced by an integral value, i.e. by the line strength S

$$\frac{1}{L}\int_{\text{line}} \ln(I_0(\nu)/I(\nu))\,\mathrm{d}\nu = nS\,. \tag{2}$$

In the case of dense or complex spectra like CF_4 or C_3F_8 definition (2) is not applicable and $\sigma(\nu)$ should be used. Since literature values with the required spectral resolution and studies on their temperature dependence above room temperature for application to plasma diagnostics were not available the absorption cross sections had to be calibrated. Following (1) the approach is straightforward for stable molecules provided L and the number density n in the sample is known. In this case σ can directly be calculated from the measured $\ln(I_0/I)$. However, due to the influence of the rapidly chirped QCL pulse the deduced absorption cross sections should be considered as effective values σ_{eff}. Typically the measured absorbance and hence σ_{eff} is underestimated in comparison to experiments without "rapid passage effect".

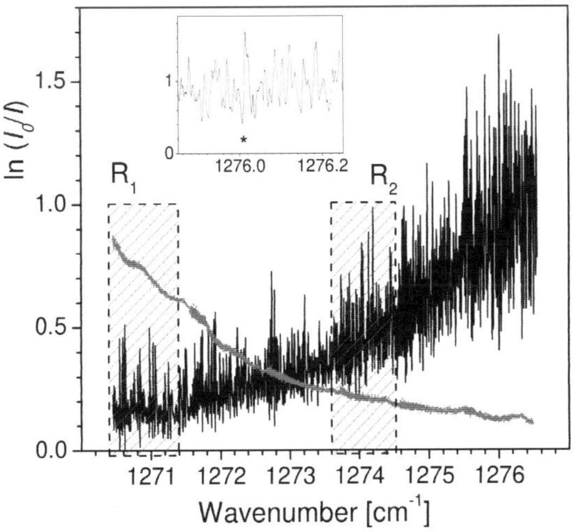

Figure 3: QCL overview spectra of CF_4 (black) and C_3F_8 (grey) measured at 10 Pa with 90 cm absorption length. The spectral micro-windows (R_1, R_2) have been defined for further analysis to discriminate absorption contributions from these two molecules. Inset: Detailed view on partly resolved CF_4 lines with indication of the "rapid passage effect" (asterisk).

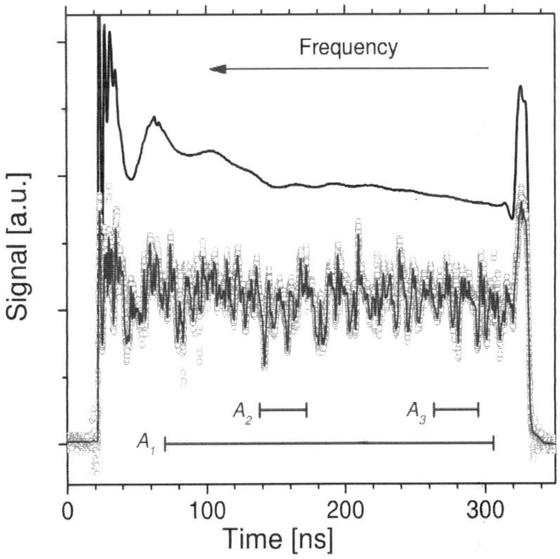

Figure 4: QCL spectra at micro-window R_2 (1274 cm^{-1}). Grey points represent single shot measurements; solid lines represent averages over 600 spectra (baseline and 70 % CF_4 at 10 Pa respectively). For the determination of σ_{eff} the three indicated ranges A_1 - A_3 have been analysed. Note: the frequency scale increases from the right to the left due to the frequency-down chirp.

As discussed in section 2 even a small inevitable jitter of less than 1 ns in the trigger may cause a non-negligible error, since not the peak but the wing of an absorption feature would be analysed at a fixed temporal position in a spectrum. This error can be markedly reduced, if not a distinct point in the spectrum is analysed, but a carefully defined range, i.e. a spectral range $\Delta\nu$ in the order of a few 0.1 cm^{-1} [9], which is in fact an additional averaging

$$\sigma_{eff}(\langle\nu\rangle) = \frac{1}{nL}\ln\left(\frac{I_0(\langle\nu\rangle)}{I(\langle\nu\rangle)}\right) = \frac{1}{nL}\ln\left(\frac{\langle I_0(\nu)\rangle\cdot\Delta\nu}{\langle I(\nu)\rangle\cdot\Delta\nu}\right) = \frac{1}{nL}\ln\left(\frac{A_0}{A}\right). \tag{3}$$

Instead of calculating the absorbance from the transmitted intensities I at ν the area (or integral) $A \approx I\Delta\nu$ can be employed in predefined ranges. For the calculation of σ_{eff} the contribution of $\Delta\nu$ cancels out in equation 3. In order to estimate the error caused by this approach three ranges A_1 - A_3 of different size have been defined for further analysis (figure 4).

The experiments to deduce σ_{eff} were carried out under flowing gas conditions at two different pressures (10 and 25 Pa) and at room temperature (296 K) in the plasma chamber. The results are shown in figure 5. For the precursor CF_4 effective absorption cross sections σ_{eff} of 4.2×10^{-18} cm^2 and 1.5×10^{-18} cm^2 are gained at 1271 cm^{-1} and 1274 cm^{-1} respectively. Conversely, 1.5×10^{-18} cm^2 and 3.4×10^{-18} cm^2, respectively, are deduced for C_3F_8. All these values should be considered as average values with a total error of about 15 %. This includes uncertainties in the pressure and gas flow measurements as well as in the data acquisition from the detector, e.g., baseline stability over time. It can be also clearly seen in figure 5 that values obtained from the fairly extended range A_1 appear between those from A_2 and A_3. Moreover, the σ_{eff} values deduced from A_2 and A_3 reproduce the general trend in the absorbance (figure 3) for all molecules, e.g., $\sigma_{eff}(A_2) > \sigma_{eff}(A_3)$ for CF_4 since A_2 was defined in all cases at higher wavenumber and thus located closer to the band origin than A_3.

3.3. Temperature dependence

The absorption cross section is typically temperature dependent which has to be considered for plasma measurements. Therefore the relative change of σ_{eff} with temperature has been studied for all ranges A_1 - A_3 in both spectral micro-windows, R_1 and R_2 (figure 6).

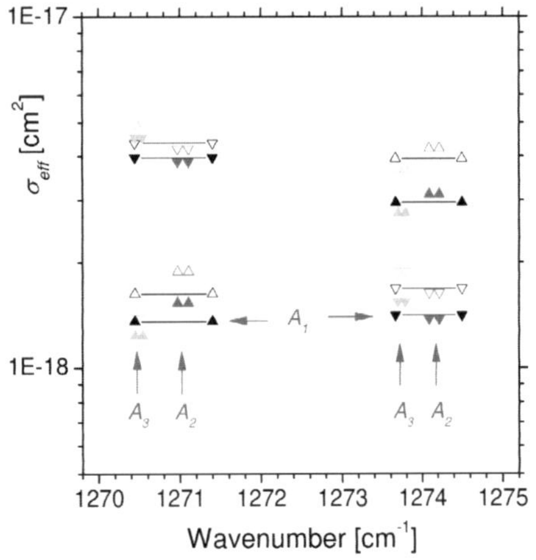

Figure 5: Calculated effective absorption cross section σ_{eff} for CF_4 (triangle up) and C_3F_8 (triangle down) at 10 Pa (open symbols) and 25 Pa (full symbols). The three analysed spectral ranges A_1 (solid line) and A_2, A_3 are indicated by arrows.

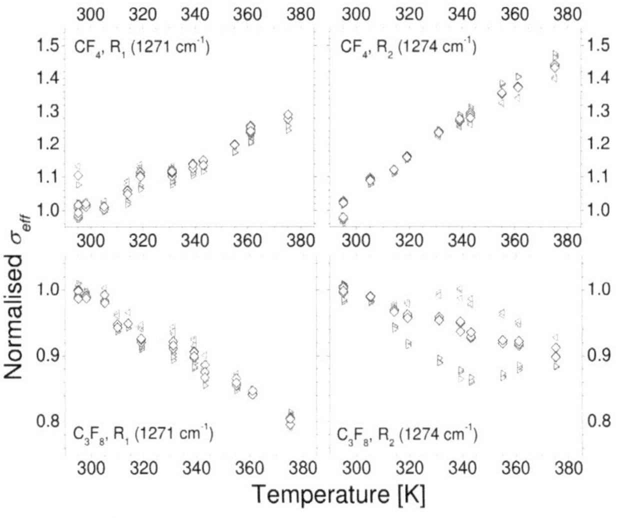

Figure 6: Relative temperature dependence of the calibrated effective absorption cross sections $\sigma_{eff}(T)$ measured at 25 Pa in a 25 cm long reference cell. All three ranges A_1 (◇), A_2 (◁), and A_3 (▷) were analysed separately. Dashed lines show a parameterisation by second order polynomials.

The measurements were performed under static conditions at 25 Pa for both molecules, CF_4 and C_3F_8. The reference cell was temperature stabilised by a heating tape which enabled temperatures up to 100 K above room temperature to be achieved. The analysis was made on average spectra (over 150 pulses) for each of the predefined ranges A_1 - A_3. Except for C_3F_8 at 1274 cm^{-1} the relative change of σ_{eff} yielded the same results at these three ranges (figure 6). For CF_4 a monotonic increase of σ_{eff} by 30 % and 50 % is observed in R_1 and R_2, respectively, for a temperature increase from room temperature to 390 K. In contrast, σ_{eff} decreases with temperature for C_3F_8, however much less pronounced than for CF_4. The decrease is less than 20 % up to 390 K. The typical error, mainly caused by uncertainties in the baseline measurements, is about 10 %. The difference in the relative temperature dependence for C_3F_8 at 1274 cm^{-1} between the micro-windows A_2 and A_3 is not clear, but might be still within the general measurement error. In order to discriminate conversion or hysteresis effects a few measurements were repeated during the cooling phase, but no significant difference was found compared to the values obtained during the heating of the cell. In order to apply the obtained temperature dependencies to the calculation of absolute number densities in forthcoming plasma measurements $\sigma_{eff}(T)$ was parameterised by second order polynomials (figure 6). Due to the monotonic behaviour an extrapolation up to 400 K or potentially 425 K might also be regarded as valid.

4. Summary

Spectroscopic studies for time resolved QCLAS measurements on fluorocarbon containing plasmas have been performed, among them the calibration of effective absorption cross sections for CF_4 and C_3F_8 at 1271 cm^{-1} and 1274 cm^{-1}. Since results of measurements with pulsed QCLs are affected by the rapidly chirped laser a careful calibration of effective absorption cross sections σ_{eff} is necessary. It was shown that defining σ_{eff} for a distinct spectral micro-window is preferable because the inherent averaging effect reduces the calibration uncertainties. Calibration in two different spectral ranges is required for the discrimination between CF_4 and C_3F_8 in a discharge. Although the determined σ_{eff} values are expected to be underestimated due to the "rapid passage effect" the order of magnitude for CF_4 ($\leq 5 \times 10^{-18}$ cm^2) stays in good agreement with non-highly resolved literature values [23,32]. It should be noted that the absolute σ_{eff} values cannot directly be transferred to other QCL systems. Every change in the chirp rate, due to a another QCL, or the bandwidth of the detection system (i.e. detector, preamplifier, digitiser) may alter the observed absorbance. Nevertheless, the normalised values still enable quantitative results to be obtained [12] or may be transferred to other systems. Furthermore, a pronounced temperature dependence of σ_{eff} was deduced for CF_4. The absorption cross section rises by up to 50 % for a 100 K temperature increase above 296 K. While the absorption cross section integrated over the entire band is temperature independent [33] the present results correlate with the tendency of σ which was found below 296 K at the current spectral position [23]. In contrast to CF_4, the temperature dependence of C_3F_8 is less pronounced and σ_{eff} is decreasing with temperature. The highly resolved C_3F_8 spectra appear as a pure drop of the baseline where no spectral feature can be recognised.

Acknowledgements

This work was supported by the German Research Foundation (DFG) within the framework of the Collaborative Research Centre Transregio 24 "Fundamentals of Complex Plasmas" (SFB-TR 24), projects B2 and B5. The authors thank D Gött, P Druckrey, U Meißner, S Saß, C Senske and F Weichbrodt for valuable technical assistance. The authors are further indebted to Alpes Lasers for their support and S Glitsch for valuable discussions.

References

[1] McManus J B, Shorter J H, Nelson D D, Zahniser M S, Glenn D E and McGoverrn R M 2008 *Appl. Phys.* B **92** 387
[2] Hay K G, Wright S, Duxbury G, Langford N 2008 *Appl. Phys.* B **90** 329
[3] Tittel F K, Bakhirkin Y, Kosterev A A and Wysocki G 2006 *Rev. Laser Eng.* **34** 275
[4] Bauer C, Sharma A K, Willer U, Burgmeier J, Braunschweig B, Schade W, Blaser S, Hvozdara L, Müller A and Holl G 2008 *Appl. Phys.* B **92** 327
[5] Miller J H, Bakhirkin Y A, Ajtai T, Tittel F K, Hill C J and Yang R Q 2006 *Appl. Phys.* B **85** 391

Journal of Physics: Conference Series **157** (2009) 012010 doi:10.1088/1742-6596/157/1/012010

[6] Tuzson B, Mohn J, Zeeman M J, Werner R A, Eugster W, Zahniser M S, Nelson D D, McManus J B and Emmenegger L 2008 *Appl. Phys.* B **92** 451
[7] McCurdy M R, Bakhirkin Y, Wysocki G, Lewicki R and Tittel F K 2007 *J. Breath Res.* **1** 014001.
[8] Cristescu S M, Persijn S T, te Lintel Hekkert S and Harren F J M 2008 *Appl. Phys.* B **92** 343
[9] Stancu G D, Lang N, Röpcke J, Reinicke M, Steinbach A and Wege S 2007 *Chem. Vap. Deposition* **13** 351
[10] Cheesman A, Smith J A, Ashfold M N R, Langford N, Wright S and Duxbury G 2006 *J. Phys. Chem.* A **110** 2821
[11] Welzel S, Gatilova L, Röpcke J and Rousseau A 2007 *Plasma Sources Sci. Technol.* **16** 822
[12] van Helden J H,. Horrocks S J and Ritchie G A D 2008 *Appl. Phys. Lett.* **92** 081506
[13] J, Capasso F, Sivco D L, Sirtori C, Hutchinson A L and Cho A 1994 *Science* **264**, 553
[14] Kazarinov R F and Suris R A 1971 *Sov. Physics Semicond.* **5** 707
[15] Beyer T, Braun M and Lambrecht A 2003 *Journ. Appl. Phys.* **93** 3158
[16] Duxbury G, Langford N, McCulloch M T and Wright S 2005 *Chem. Soc. Rev.* **34** 921
[17] McCulloch M T, Duxbury and Langford N 2006 *Mol. Phys.* **104** 2767
[18] Haverlag M, Stoffels W W, Stoffels E, den Boer J H W G, Kroesen G M W and de Hoog F J 1996 *Plasma Sources Sci. Technol.* **4** 260
[19] Haverlag M, Stoffels E, Stoffels W W, Kroesen G M W and de Hoog F J 1996 *J. Vac. Sci. Technol.* A **14** 380
[20] Gabriel O, Stepanov S and Meichsner J 2007 *J. Phys.* D: *Appl. Phys.* **40** 7383
[21] Gabriel O, Stepanov S, Pfafferott M and Meichsner J 2006 *Plasma Sources Sci. Technol.* **15** 858
[22] NIST Chemistry WebBook (URL = http://webbook.nist.gov/chemistry)
[23] McDaniel A H, Cantrell C A, Davidson J A, Shetter R E and Calvert J G 1991 *Journ. Atm. Chemsitry* **12** 211
[24] Hurley M D, Wallington T J, Buchanan G A, Gohar L K, Marston G and Shine K P 2005 *Journ. Geophys. Res.* **110** D02102
[25] Poll H U, Hinze D and Schlemm H 1982 *Appl. Spectrosc.*. **36** 445
[26] Suzuki T, Okada H and Fujiyama T 1979 *Bull. Chem. Soc. Japan* **52** 2505
[27] Takami M 1981 *J. Chem. Phys.* **74** 4276
[28] Höjer S and May R D 1996 *J. Mol. Spec.* **178** 139
[29] Stepanov S, Welzel S, Röpcke J, and Meichsner J 2008 *J.Phys.: Conf. Ser.* submitted
[30] Yamada C and Hirota E 1983 *J. Chem. Phys.* **78** 1703
[31] Ballard J, Knight R J and Newnham D A 2000 *J. Quant. Spectrosc. Rad. Transf.* **66** 199
[32] Cruden B A, Rao M V V S, Sharma S P and Meyyappan M 2002 *Plasma Sources Sci. Technol.* **11** 77
[33] Nemtchinov V and Varanasi P 2003 *J. Quant. Spectrosc. Rad. Transf.* **82** 461

AUTHOR INDEX